Nonlinear Psychophysical Dynamics

Robert A.M. Gregson
The University of New England
New South Wales
Australia

 Routledge
Taylor & Francis Group

NEW YORK AND LONDON

First published 1988 by Lawrence Erlbaum Associates, Inc.

This edition published 2011 by Routledge
605 Third Avenue, New York, NY 10017
4 Park Square, Milton Park, Abingdon, Oxon OX14 4RN

Routledge is an imprint of the Taylor & Francis Group, an informa business

Library of Congress Cataloging in Publication Data

Gregson, R. A. M. (Robert Anthony Mills), 1928-
    Nonlinear psychophysical dynamics / Robert A.M. Gregson.
       p.   cm.
    Bibliography: p.
    Includes index.
    ISBN 0-8058-0015-8 :
    1. Psychophysics--Mathematical models.   2. Nonlinear theories.
 I. Title.
 BF237.G74 1988
 152.8--dc19   88-12109
                                                                        CIP

ISBN 13: 978-0-8058-0015-9 (hbk)

Preface

This set of lecture notes grew out of work done at various times between 1984 and 1987, first in the German Federal Republic, and later and mostly in Australia. There is now a tradition in mathematics and the physical sciences of publishing work in its earlier formative phases in so-called lecture notes; this practice has not spread to psychology but the content of this particular monograph makes it apposite. Nothing final and definitive is being asserted, but a deliberate break with a tradition going back to before Fechner's foundation work in 1860 is made.

Put oversimply, the major tradition in psychophysics is one in which an environment is postulated to generate, deterministically, a random series of events. These become stimuli for a probabilistic error-loaded organism, whose overall (i.e. externally observable) stimulus-response relationships are approximated by linear models, with superimposed Gaussian residual noise. The whole endeavour leads reasonably into psychophysical scaling, and the problems that consequently follow stem from trying to establish, both experimentally and mathematically, the regions within which the organism generates outputs that have metric properties, given that some relevant properties of the physical world may be summarised in the same framework.

This monograph quite deliberately turns the whole exercise on its head. The organism, in so far as it constitutes a single sensory channel with one-dimensional inputs, is initially treated as a nonlinear deterministic process, within a stochastic environment. Gaussian fuzziness is pushed to the outside. The mathematical assumptions used as a skeletal representation of a sensory channel are based on a difference equation of a type whose properties have only been explored in depth since the 1970's. The Fechnerian tradition in contradistinction goes back to Gauss and Laplace at the end of the 18th century. The problems of identifying when a nonlinear model fits data and when it does not are quite different from those of the more familiar general linear model, and are consequently treated *de novo* here. The use of time series analyses of system input-output transfer functions as a means towards identification of internal system dynamics, particularly of distinguishing random behaviour from chaotic behaviour, is noted. This is not a how-to-do-it cookbook, but the relation to earlier work by this author and others which is built on is cited, specifically in Chapters 9 and 12 where new real data are used illustratively.

Because the mathematical apparatus here is different from, and owes little to, the regression equations first introduced by Fechner, and deriving from Gauss's work in astronomy as the method of least squared residuals, it is in principle possible to treat the topic as though the tradition of classical psychophysical equations does not exist, but only that a mountain of detailed experimental data, waiting to be given some coherence and structure, lies in an archive going back a century or more. Hence, we treat the topic not as a comparative analysis of classical and new psychophysics, but as a self-contained exercise in modelling the properties of reported data, "warts and all", from a fresh starting point. Classical psychophysical models have been described repeatedly, in innumerable textbooks, and the reader who wishes to compare their structure and efficiency with what is done here should pursue the sources in parallel. This text is solely about one class of models, just as other texts are solely about previous models, such as power laws or signal detection.

A final chapter, 14, summarises what is done and what are outstanding tasks. Psychophysics is in a scrappy state, still lacking the analogue of a periodic table or an atomic theory, and the value of any approach rests on it being able to tie together what is apparently diverse and unrelated with one model of minimum complexity. In this study complex variables are intentionally introduced and used, which is not usual in psychophysics. It is therefore historically interesting to note that already in 1801 Gauss had introduced complex numbers in paragraph 337 of **Disquisitiones Arithmeticae**, and done it without finding it necessary to give definitions or prior justification. Yet agonising over the legitimacy of using complex variables in the physical and biological sciences went on for subsequent decades. They are introduced here for two reasons: to give particular dynamics to a model, and to give a basis for a potential split of activity into two levels, which can be thought of as analogues of the more familiar signal-noise dichotomy which is made, but in real variables, in more traditional thinking. The point of using a simple non-linear equation in complex variables is that amazingly diverse and intuitively inaccessible properties follow sometimes, but only sometimes, from very small changes in parameter values. Instead of creating separate models for different levels of stimulus intensity and discriminability, changes in the few parameters of one model should suffice. In such an approach it becomes critically important to show what constraints are necessary on parameter manipulations in order to make sense. A model that fits anything is not falsifiable, and hence empty of meaning. Space must therefore also be devoted to showing how to go wrong.

So, most of the monograph is taken up with exploring some of the properties of a single related family of models. There is only one basic equation, [2.2], and the rest derives by imposing simple constraints on its parameters, or more interestingly on the mapping of environmental properties onto its parameters. Nothing is assumed of any subtlety about the mapping of system output into observable responses; this latter topic is strictly one in cognitive psychology, and needs additional assumptions not made or discussed here.

The fact that dynamic processes can move between stable but not necessarily equilibrium phases, and can create their own internal chaos which to an outside observer may be indistinguishable from random noise, unless an appropriate analysis of stimulus-response relations is conducted in the time domain, has emerged in mathematics, in the physical sciences, and in physiology. It has equally cogent relevance in psychophysics.

The motivation for pursuing the path taken here was one of "but what happens if, instead....?" on the theoretical side, in the middle of trying to reconcile some physiology with psychology, and on the experimental side of trying to tie together phenomena which arose in the author's olfactory psychophysics laboratory. Using traditional methods of data analysis is often a case of using filters to destroy information, not one of reporting the fine structure of behaviour. What is noise to one model is structure to another more sensitive one. For technical reasons most of the mathematics displaying the structure of dynamic systems has arisen in those areas of physics which have not much impinged on psychology.

A psychophysics which uses 20th century instrumentation, such as on-line recording and contingent sequence-control in stimulus presentation, and marries these with 20th century abstract systems theory, shouldn't be expected to continue to resemble what was in effect a marriage of brass instruments and linear equations in the mid 19th century.

I would like to thank many people for help and encouragment. In Germany, where in 1984 I was supported by Deutsche Forschungsgemeinschaft, Prof. Dr. C. H. Micko of Technische Universität Braunschweig for hospitality, and Prof. Dr. V. Sarris of J. W. Goethe Universität Frankfurt for discussions on transposition phenomena. In Australia my research assistants Ian Price and Fiona MacBride were cheerful, reliable and patient colleagues, funded by the Australian Research Grants Scheme in 1986 - 1987. Ian Ellis converted me to TEXand made a lot possible in reasonable time. Computers used were PDP11/23,40,70, VAX, DEC20, with original FORTRAN programs and subroutines from IMSL, NAGF, IGS, PLOT79, and the SCA system for time series analyses. My wife Diana has come to recognise my processes of literary gestation, having lived through two books already; for her tolerance and support I am again grateful. Lawrence Erlbaum insists this is really a book, which is perhaps more flattery than I deserve.

Seminars I have given at various universities in Europe and in Australia have provided some feedback and an understanding of why this approach produces initial bafflement and then curiosity in somebody with only a traditional upbringing in either experimental or mathematical psychology. Pandora once opened a large pot and let many mischiefs into the world, but that step is one which must be taken repeatedly in opening any blackbox of a biological system.

"Ητοι μὲν πρώτιστα Χάος γένετ·´

αὐτὰρ ἔπειτα Γαῖ᾽εὐρύστερνος

ΘΕΟΓΟΝΙΑ, 116 - 117.

πλείη μὲν γὰρ γαῖα κακῶν, πλείη δὲ θάλασσα·

νοῦσοι δ᾽ ἀνθρώποισιν ἐφήμέρη, αἰ δ᾽ἐπὶ νυκτὶ

αὐτόματοι φοιτῶσι κακὰ θνμτοῖσι φέρουσαι

σιγῇ,

ΕΡΓΑ ΚΑΙ ΗΜΕΡΑΙ, 101 - 104.

Hesiod, circa 550 B.C.

Contents

Contents

1 General Qualitative Dynamics of some Nonlinear Systems

Nonlinear systems have become a focus of intense and extensive intellectual activity, not only in physics but by derivation in the biological sciences (Schuster, 1984, Hao, 1984, Holden, 1986). An approach to modelling complex and traditionally intractable processes of such generality obviously can be the subject of an overview from a very wide perspective; this has already been done very competently by the authors just cited, and is not our objective here. Rather, within the limited framework of a nonlinear dynamical systems theory approach, we concentrate upon only one topic, the generation of sensory intensity as a response to a physically varying environment.

This approach is a sort of theoretical psychophysics, in which a transfer function representation is written without regard to the specific mechanisms of sensory transduction in a given modality such as vision, hearing, touch, smell, or whatever. It is in contradistinction to the approach taken, for example, by Laming (1986), which focusses on stochastic models of the initial stage in the perceptual process, and so we are here closer to what he has called Sensory Judgment as opposed to Sensory Analysis. Whether the two can be so readily demarcated is not a settled matter, but this author prefers to think of the two approaches as complementary. Most detailed models in psychophysics are very properly concerned with what may be called[1] (a) the microfunctional representation of the signal conversion from environmental to neural events, (b) the series-parallel transmission of signal sequences in neural pathways from receptors to higher brain centres, and (c)

---

[1] I am indebted to Prof. Dr. U. Mortensen for fruitful discussions on this topic.

the selection of output representations (responses) of some of the properties of the neural activity stored in the higher centres.

Our focus here is heavily on (c), and we choose to telescope (a) and (b) into a single *one* $\Longrightarrow$ *one* or a related noisy mapping of input onto a central state variable. Thus the interest is in the mass action of sensory systems that exhibit some input-output invariances under restricted but identifiable stimulus conditions. These invariances are not necessarily simple, but may be represented only in the *structure* of nonlinear difference or recursion equations, whereas the *parameters* reflect in their variability the current condition of the observer as a system.

It is perhaps easiest to begin with a relatively familiar result and work away from it. Traditionally the zero-order (*i.e. time independent*) relationship between physical stimulus magnitudes and psychological response intensity is represented as a psychometric function which is

        (a) continuous and everywhere single-valued

        (b) bounded both above and below

        (c) monotone increasing between the bounds

        (d) approximated by a cumulative normal ogive (CNO)

This picture is found in the older textbooks, well-established by the first decade of the 20th century, with some second-order modifications (see Guilford, 1954, for an extensive review of such early curve-fitting) . The ubiquitous normal curve assumption of course pervades the models of Thurstone (1927) and again surfaces in the more usual, but not necessary (Egan, 1975) assumptions of signal detection theory.

Nothwithstanding the very good fits that psychophysical data can yield when plotted as a straight line on log-probability paper, given some meticulousness in controlling the conditions of data collection, repeated attempts do surface from time to time to derive the psychometric function from other process assumptions (e.g. Freeman, 1975, and a symposium in the British Journal of Mathematical and Statistical Psychology, 1984). A possible rationale for accepting the CNO form stems from the idea that sensation is mediated by a vast number of independent elements or channels, and a progressive temporary recruitment from the population operates as the level of input to the system is increased, so that more and more elements are incorporated in the process or channel to carry stronger and stronger signals. The weakness of this conceptualisation quite simply lies in the notion that neural substrate mass action is appropriately modelled by a vast aggregate of independent elements; in fact the channels are in a network and the minimal connectivity of such a network can be shown to support stable dynamic properties which a random collection could not display (Peretto and Niez, 1986a,b).

The interest in locally connected networks, where each element has some interaction with its neighbours, and interaction is not necessarily inversely proportional to the functional (*not* literal physical) distances between elements, lies in various properties. For example:

(1) The dynamics of large aggregates in connected networks are independent of the number of elements involved; they do not increase in complexity with the size of the network.

(2) The number of patterns of nonlinear stability and instability, such as catastrophes, limit cycles, hysteresis, is limited and in many cases they can consequently be indentified and modelled.

(3) Large aggregates may function effectively as single recursive processes, and exhibit characteristics of a single feedback loop, though probably with a complicated (high order) transfer function.

All this means that large collections of neurons, which are the substrate of any sensory pathway to consciousness, have their own level of dynamics which are not derivable just by adding up the activity of their components. The behaviour of a collection of idealized cells, each with inhibitory and excitatory inputs and outputs, is not something like a complicated telephone exchange. A network or aggregate of interconnected cells has emergent properties which can be at one moment very simple, at another very complicated, depending upon the level of input and the subsequent rate of energy dissipation. Fortuitously, when such behaviour is simple it almost looks like traditional psychophysics.

Previous examples suggest that there are two main ways to go in exploring a nonlinear psychophysics which can potentially relate back into the modelling of the related neurophysiological substrates at the mass action level; either a recursive complex difference equation of order 3 or above, or three or more linear difference equations (in 3 variables) coupled to each other. The second approach includes the Lorenz (1963) equations which have been the starting-point for many analyses and simulations of physical systems. However, in psychophysics, as distinct from neurophysiology, there is rarely sufficient justification for using systems of coupled equations; we do not know enough about a sensory system to justify such strong hypotheses. It is thus expedient to sketch out the difference between psychophysics and neurophysiology a bit further.

Although psychophysical models which are typically very simple linear input-output equations are written without specifying that they depend on the choice of units and the sensory dimension involved, obviously their solution involves fixing parameter values which are characteristic of the dimension within the modality which is being studied. Even if the representation of sensory intensity is supposed to have some general common basis extending across modalities, which makes it possible to write of and conduct experiments on cross-modal matching of apparently qualitatively diverse dimensions of sensation, such as loudness and stickiness (!!!), as soon as we move to the very detailed study of a specific modality the dominance of questions relating neurophysiological activity to experienced sensation quality and intensity over any very general formalization is inevitable.

The extensive and detailed models of psychophysics are almost all specific to vision or to hearing, and it is consequently mainly in those two modalities that there exist data bases, derived with separate experimental methodologies, that are sufficiently firmly based in detailed parameter estimation for it to be possible to derive theoretical psychophysical input-output topologies from neurophysiological assumptions and data. Proceeding in the reverse direction is not possible because psychophysical relations are overall input-output gain functions and so cannot logically entail tight constraints on the neurophysical substrate's functional equations. There are many ways internally of achieving the external properties of a black box. One may express this redundancy by saying that there is a many$\Longrightarrow$one mapping from X onto Y, where

$$\left\{ \begin{array}{l} X_{j+1} = A_j \cdot X_j + U_j + w_j \\ Y_{j+1} = B_{j+1} \cdot X_{j+1} + v_j \end{array} \right\} \qquad [1.1]$$

in state space notation; the U_j are the stimuli on trial j, the X_j, X_{j+1} are the central processes serving as state variables and the Y_{j+1} are the observable psychophysical outputs or quantified responses. The w_j and v_j are noise, no constraining assumptions are made about $var(w)$, $var(v)$, $covar(w, v)$, or their autoregressions. Putting suffices to A and B indicates that these operators are themselves potentially time-dependent. Obviously [1.1] is too loose to serve as more than a heuristic, but starting with a generalised nonlinear equation of Volterra form is a comparable exercise which has been pursued (an der Heiden, 1980). In psychophysics a system theory approach which treats some formal representation of the (implicit, unobservable) dynamics of sensation as state variables, to be estimated given some constraints on model structure, is viable.

The properties which neural networks have, as a consequence of their interconnectedness, are exceedingly complex, and suggest that we should expect (as we do indeed get), multistability, and sometimes apparent instability in the time series representations of output. A network has to be regarded as both transmitting signals cleanly (as though an almost noiseless path has been cut through it), or diffusing the input into an excitation pattern in space and time with high entropy, and perhaps going into reverberations. The most interesting developments in nonlinear systems theory suggest gross analogies between energy transmission in large aggregates of molecules (e.g. the Belousov-Zhabotinsky reaction) and signal transmission in neural assemblies. In both cases the number of "single elements" is 10^8 or more and the behaviour of single elements is not observable.

The axiomatic structures of complex models of sensory systems commonly presuppose that the distributions of events would be gaussian, and, or, that they would be perturbed by gaussian noise. That is, any transmission process may show moment-to-moment variability in the magnitude of its outputs, and this may be treated as the consequence of the output being the result of a large number of partially uncoupled system elements, each probabilistic in time in its transduction properties.

This assumption of gaussian noise and/or intrinsic variability breaks down if advanced as an unqualified generalisation, because output distributions show statistical forms markedly deviant from the normal distribution. When this happens special additional assumptions about behaviour under limiting cases, such as responding to very weak stimuli , have to be made if we are to retain the assumptions; observed output distributions might then be treated as mixtures of simpler probability functions.

There is yet another problem, which arises from the restriction which may be imposed on the human observer to report only "yes/no" or binary analogues such as "lesser/greater" . If the organism is only capable of detection/non-detection or discrimination/non-discrimination, then the continuous variability of processes within the system has to be degraded in some way when an output is selected by the observer. Commonly some gating mechanism is assumed to operate, if the internal variable X is above a threshold X_c then the output Y is observed. X_c is not itself observable, but is a construct which can be given a stochastic variability in time, with whatever autoregressive structure that seems plausible. Of itself such a structure may be insufficient to capture observed output distribution properties, and so some convolution or integration of the X variable to give

$$X'_j = \sum_{k=0}^{m} v_k \cdot X_{j-k} \qquad [1.2]$$

and then use X'_j as the variable for the stochastic inequality $X'_j \neq X_c$, with

$$\left\{ \begin{array}{ll} Y_j = 1 & \text{iff} \quad X'_j > X_c \\ Y_j = 0 & \text{iff} \quad X'_j \leq X_c \end{array} \right\} \qquad [1.3]$$

as the output decision rule, is advanced.

The consequence of introducing threshold mechanisms like [1.3] is of course to create nonlinearity, which increases the *a priori* probability that the model would fit behaviour, because psychological systems are typically nonlinear. Introduction of any stochastic gating like [1.3] into a quasi-continuous system may be used to generate point processes as output series. If an experiment being modelled is constructed in such a way that the outputs are only point processes (such as repeated line crossings between two states) then the modelling is reduced to matching the distribution statistics of theoretical and observed point processes under a range of input conditions.

These comments on what is actually a large body of theory building in psychology are deliberately terse, precisely because the intention here is to start from a different basis, and hope to proceed from a deterministic recursive dynamics to generate apparently stochastic outputs, the autoregression of outputs, and the system nonlinearities, from the one same algebraic source. This does not exclude gaussian noise as a context variable for the system modelled, but it alters fundamentally the status of any stochastic part of the system; it moves it, as it were, to the periphery, or more strictly speaking into the environment. A biological system is locally entropy decreasing, and there is thus some legitimacy in considering models which do not add significantly to the entropy of the input. This, incidentally, suggests additional statistical tests on the comparison of the input and output entropies. A system which degrades a continuously varying input to a point process output obviously loses information, but if we allow the output also to vary almost continuously, then the information transmitted may be greater, but with some entropy loss.

The choice between modelling observed behavioural output as binary or as almost continuous is pervasive in experimental psychology; it is for example the dichotomy between classical psychophysics and scaling by direct magnitude estimation. At one extreme the theorist eschews anything but binary outputs because the metric properties of a quasi- continuous observed numerical response are indeterminate, at the other extreme any and all response information that the subject generates is treated as legitimate material for modelling and is approximated by functions with metric properties.

The nonlinear system Γ to be studied here usually generates output distributions, for a range of inputs, which are not representable by continuous functions but which are locally dense and, under some subregion of the system parameter space, can produce input-output gain functions resembling the ubiquitous psychometric ogival forms.

If, however, we want to consider the output as a point process, when the degradation of output is created by the system and not *subsequently* by an external observer, then generally but not always it would be necessary to add a gating mechanism as in the stochastic case. Because the internal output of the recursive nonlinear process, Y_{obs}, is already variable (*one* $\Longrightarrow$

many) with respect to the input U, so [1.1] could be applied without adding any assumptions about the variation in time of some Y_c. This we have done, for example, to produce an analogue of some signal detection processes.

There is perhaps a reason why we may wish to degrade Y_{ob}, to a point process Y', and then work with probabilities based upon the distribution of Y'. This would be done to replicate the form of results produced by psychophysical experiments on thresholds (of various sorts, including, say, flicker fusion). If this course is pursued, the recursive nonlinear process in Γ (as used here in Chapter 2) is essentially performing the same role within the system model as the gaussian variance generator process does in a stochastic model; it is the unobservability of differences between finite samples from a system in deterministic chaos and one in stochastic perturbation which in a restricted sense makes the two approaches interchangeable. As the time series properties of Y_{ob}, and hence of Y' will however not be the same as the Y based on X' in [1.1] the models could have identifiably different properties in respect of their autoregressive structures.

Limit Cycles, Strange Attractors and Deterministic Chaos.

We have already mentioned without definition some properties of nonlinear systems in the previous paragraph; it is now necessary to expand and explain those comments. Biological systems operate at least a substantial proportion of their time away from equilibrium; one might even think of behaviour as an intermittent directed effort to return to equilibrium (Helson provided us with the obvious example of a psychologist focussing on this as a central theme in describing behaviour) and to restabilize at new input-output levels. The idea is quite old in physiological psychology, dating back to the early 1900's and to Cannon's *wisdom of the body.*

Increasingly systems theory has concerned itself with the behaviour of living or mechanical entities operating away from equilibrium. Instead of either creating in the laboratory, or selectively observing in the outside world, the average input-output relationships of a body of data taken out of its temporal context, and then necessarily using central tendency statistics as a filter to partial out noise, the interest shifts to the sequential dynamics of a system when some conditions are observed. These conditions are:

   (i) the system is not in static equilibrium; it does not absorb onto a single output from a given input,

   (ii) the system is locally entropy reducing,

   (iii) the system as a whole is dissipative, it uses energy and is irreversible,

   (iv) the system is quasi-closed; it can be approximately treated as closed only with respect to a restricted set of variables in a short time span,

(v) small changes in input do not necessarily lead only to small changes in output,

(vi) the system is strongly dependent upon its initial conditions.

Conditions (i) through (vi) are dynamic characteristics of a diversity of physical and biological processes which have become the subject of intensive mathematical and experimental analysis (Hao, 1984, West, 1985). Fascinating links between the dynamics of complex systems and the aesthetics of their graphical representation are displayed in the work of Peitgen and Richter (1986), which should be consulted not only for the beauty and elegance of the presentation, but also for an illuminating essay by Mandelbrot which pointedly records the fruitfulness of sometimes taking a maverick position in science and in mathematics. The question of central interest is to see heuristically how we might use results concerning invariant dynamic properties which have been identified as associated in a very general, even qualitative, sense with the above conditions. Precisely what *identified in a general sense* means is in fact a question of technical complexity and subtlety; it is the core problem in the abstract mathematics of nonlinear systems which has to be addressed before we can with any confidence predict the grosser qualitative dynamics which will most probably be observed in the real world. There is increasing acceptance that biological systems, within which class are included higher brain functions, are only meaningfully represented in nonlinear forms (Bienenstock, 1986). The intractability of nonlinear mathematics, and the success of linear approximations in classical physics and its derived sciences, which served as models in a powerful way for psychology, have both been obstacles to taking the view that nonlinear dynamics might give us powerful and rigorous formal insights into human behaviour. The gap between subcultures has been looked across by some workers, even tentatively bridged locally (McFarlane on control and motivation, 1974) but is usually treated as hopelessly wide. A new explosion of knowledge encourages us to try again.

A recurrent feature of nonlinear systems is their capacity to go into quasi-cyclic dynamic patterns when their parameters are away from the zone of the parameter space which defines equilibrium. Such patterns appear, unless the analysis for their identification is adequate, like weakly-coupled responses to periodic inputs, or they resemble random noise. They can occur and be separately detected in the presence of noise, provided that the noise itself is given an adequate statistical definition.

The idea of a strange attractor as an identifiable form of dynamic stability came into system theory about a century after the precursor idea of limit cycles. A limit cycle is a reasonably obvious and recognizable pattern, because the observable output of such a system is then quasi-cyclic and can be approximated by one or more frequencies in a Fourier spectral analysis or an autoregressive spectrum; it looks like the direct response to a periodic (fluctuating) input but in fact it is autonomous and internally generated, persisting when the environment is in a steady aperiodic state. Neurophysiological examples are well known, for example the immune responses (Pimbley, 1973) and endocrine systems (Abraham, Kocak and Smith, 1985) and in the olfactory cortex (Freeman, 1975, 1983, Freeman and Viana di Prisco, 1986) as well as in related psychopathology over a longer time scale (King, Barchas and Huberman, 1983). An example of limit cycling in continuous variable models of human tracking rhythms has been given by Kay, Kelso, Saltzman and Schöner (1987).

The notion of a strange attractor lacks a generally used formal definition, but can be sufficiently characterised by some of its properties in the phase space of the system. That is to say, we have to plot the output against either its own values lagged in time, or against its rate of change, and then there results an endless trajectory which is dense inside one region of the space, but never quite touches the boundaries of that region, and eventually passes through all the points in that region. The location, or shape, of this attractor may be a torus (like a doughnut) and the system meanders forever inside it. The spectral analysis of a strange attractor may be a broad band of frequencies, or white noise.

The dimensionality of an attractor is always less than the dimensionality of the system, and can be non-integer. The concept of non-integer dimensionality needs special definition and the reader is referred to Schuster (1984) on this point. For example (West, 1985), in a 3-dimensional system one can have a 2-dimensional attractor which has no dominant frequency but energy at all frequencies in one dimension, and a broad spectrum *plus* isolated peaks on the other dimension. The autocorrelation of the output will thus, on the second dimension, be the sum of a nonperiodic component decaying to zero and a periodic component that does not decay. This pattern is found in biological processes, and in human psychophysics (Gregson, 1984). It is the peculiar mix of broad spectrum and specific frequency activity that characterises strange attractors (which are virtually a synonym for deterministic chaos), and in the work of Nicolis (1983) has facilitated a new interpretation of the electroencephalogram. West (1985) notes that a system which is both dissipative *and* open to the environment would probably, and not merely possibly, have one or more strange attractors. By definition a system which has a strange attractor in it is said to be chaotic; this is the particular sense in which deterministic chaos is meant when we use the term to describe behaviour.

With these comments it is worth quoting[1] Schuster's (1984, p.93) defi-

---

[1] It is important to note that at this time (1988) there is no generally

nition of a strange attractor as *(a) a bounded region of phase space to which all sufficiently close trajectories from the so-called basin of attraction are attracted asymptotically for long enough time. The basin of attraction can have a very complicated structure. Furthermore the attractor itself should be undecomposable, i.e. the trajectory should visit every point on the attractor in the course of time. (b) Strange attractors have sensitive dependence upon initial conditions. (c) To describe a system the attractor has to be structurally stable and generic – a small change in the parameters of a (first order) differential equation for a dissipative system changes the structure of the attractor continuously.*

The degree of chaos in a system can be defined as K(-entropy); the rate at which information about the state of the system is lost in the course of time. K-entropy is inversely proportional to the time interval over which the state of the system can be predicted. K is greater than zero for unidimensional chaotic motion.

Whilst patterns of chaotic behaviour are ubiquitous, but not necessary, in nonlinear systems, their precise form and generation vary considerably with the structure of the system involved, and a sufficiency of complexities and exceptions exist for us to be warned to explore **any** equation *that is not already well understood* as new uncharted territory. Such exploration can be analytic, by simulation, or both.

In particular, though there exist many studies (Garrido and Simo, 1983) of three-dimensional coupled linear systems (starting from the Lorenz, 1963 model) and of one-dimensional mappings onto the unit interval, relatively little is known empirically, and even less in a strictly analytic sense, about either sort of model with complex variables. A complex extension of the Lorenz equations (Fowler, Gibbon and McGuinness, 1982, 1983) has been studied and revealed interesting properties according to the degree of internal diffusion in the system. The introduction of complex variables instead of real variables can represent some type of noise perturbation; whether or not such noise affects the qualitative dynamics which are observed as the crucial system parameters are varied is not a simply answered question. In some circumstances noise may leave the dynamics unchanged, in other cases the random noise may be inextricably confounded with deterministic chaos (Abraham, 1985).

Pursuit of the topology of abstract mathematical systems gives us unfortunately insufficient insight into how real systems, or analytically intractable models chosen for their simulation plausibility, can behave. Instead we need to implement them, even simulating by using computer

---

accepted definition of a strange attractor, but that we can get fairly close to one by specifying its properties in contradistinction to other attractors. See Parker and Chua (1987).

graphics and video films. This point has been made repeatedly in the study of chaotic systems, for example by Peitgren and Richter (1986). But the richness of nonlinear dynamics and the existence of transient complex phenomena, such as strange attractors and dimensional catastrophes, is well known and has produced a diversity of measures such as the already-cited K-entropy, or Lyapunov indices, and has led to a suggestion that the time series analysis of outputs (Packard, Crutchfield, Farmer and Shaw, 1980, Guckenheimer, 1982, Grassberger and Procaccia, 1983) would be sufficient to identify when a system is in a quasi-periodic phase, and in contrast when it is in chaos.

This suggestion about the adequacy of time series analysis as a sufficient means of identification is disputed, because counterinstances can be constructed (Garrido and Simo, 1983, p. 7, Takens, 1981). However in the context of studying a system whose abstract identification has been sufficiently narrowed down, it can be useful, and one would routinely adopt it as one of a range of methods to check the compatibility of data with tentatively identified system models, each such method being of itself necessary but not sufficient.

From the generality of some phenomena in nonlinear dynamics we can predict some changes in the behaviour of a system as control parameters are increased; for example we almost always (the exception noted by Tsuchiya, 1984 is of interest) expect abrupt changes in the qualitative form of output series for given input series which are random or fixed in their autoregressive structure, as parameters which are effectively only gain constants are increased in the one-dimensional family of recursive systems in discrete time.

Though the three-dimensional examples have some potential application in biopsychological systems, such as are implicated in neurohormonal disturbances (Abraham, Kocak and Smith, 1985) in psychophysics it seems wisest to eschew such complicated systems. In the physical sciences carefully generated conjectures concerning the plausible form of connected differential equations for quasi-closed systems are possible from a long traditon of experiment and classical mathematical analysis, but the same cannot yet be said for most of psychology.

If a nonlinear deterministic model is advanced it must, to be seriously considered as an alternative, be able to reproduce well-established empirical results as well as to make predictions which in their general form or in their fine detail were not derivable from previous models. This follows from what is called the *Principle of Correspondence.* (Gregson, 1983, p. 405). There is, however, no obligation to reproduce the precise predictions of any previous stochastic model unless it can be shown that two models, stochastic and nonlinear deterministic, are equivalent in their accuracy and are both valid up to the resolution of measurement of observations that is possible in pertinent actual experiments.

References

Abraham, R. H. (1985) Is there Chaos without Noise ? *In* Fischer, P., and Smith, W. R. (Eds.) *Chaos, Fractals and Dynamics.* Lecture Notes in Pure and Applied Mathematics Vol. 98. New York: Marcel Dekker.

Abraham, R. H., Kocak, H., and Smith, W. R. (1985) Chaos and Intermittency in an Endocrine System Model. *In* Fischer, P., and Smith, W. R. (Eds.) *Chaos, Fractals and Dynamics. Lecture Notes in Pure and Applied Mathematics Vol 98.* New York: Marcel Dekker.

Bienenstock, E. (1986) (Ed.) *Disordered Systems and Biological Organization.* NATO ASI Series, Vol. F20. Berlin: Springer-Verlag.

Egan, J. P. (1975) *Signal Detection Theory and ROC Analysis.* New York: Academic Press.

Fowler, A. C., Gibbon, J. D., and McGuinness, M. J. (1982) The complex Lorenz equations. *Physica, 4D,* 139 - 163.

Fowler, A. C., Gibbon, J. D. and McGuinness, M. J. (1983) The real and complex Lorenz equations and their relevance to physical systems. *Physica, 7D,* 126 - 134.

Freeman, W. J. (1975) *Mass Action in the Nervous System.* New York: Academic Press.

Freeman, W. J. (1983) Dynamics of Image Formation by Nerve Cell Assemblies. *In* Basar, E., Flohr, H., Haken, H., and Mandell, A. J. *Synergetics of the Brain.* Berlin: Springer-Verlag. pp 102 - 121.

Freeman, W. J., and Viana di Prisco, G. (1986) EEG Spatial Pattern Differences with Discriminated Odors Manifest Chaotic and Limit Cycle Atractors in Olfactory Bulb of Rabbits. *In* Palm. G., and Aertsen, A. (Eds.) *Brain Theory ,* pp. 97 - 120. Berlin: Springer-Verlag.

Garrido, L., and Simo, C. (1983) Some Ideas about Strange Attractors. *In* Garrido, L. (Ed.), *Dynamical Systems and Chaos.* Lecture Notes in Physics. Berlin: Springer-Verlag.

Grassberger, P., and Procaccia, I. (1983) Measuring the Strangeness of Strange Attractors. *Physica, 9D,* 189 - 208.

Gregson, R. A. M. (1983) *Time Series in Psychology .* Hillsdale, New Jersey: L. Erlbaum Associates.

Gregson, R. A. M. (1984) Invariance in time series representations of 2-input 2-output psychophysical experiments. *The British Journal of Mathematical and Statistical Psychology, 37,* 100- 121.

Guckenheimer, J. (1982) Noise in chaotic systems. *Nature, 298,* 358 - 361.

Guilford, J. P. (1954) *Psychometric Methods,* New York: McGraw-Hill.

an der Heiden, J. (1980) *Analysis of Neural Networks.* Lecture Notes in Biomathematics No. 35. Berlin: Springer-Verlag.

Hao, B.- L. (1984) *Chaos.* Singapore: World Scientific Publishing Co.

Holden, H. V. (Ed.) (1986) Chaos. Princeton: Princeton University Press.

Kay, B. A., Kelso, J. A. S., Saltzman, E. L., and Schöner. G. (1987) Space-Time Behavior of Single and Bimanual Rhythmical Movements: Data and Limit Cycle Model. *Journal of Experimental Psychology: Human Perception and Performance, 13,* 178 - 192.

Laming. D. (1986) *Sensory Analysis.* London: Academic Press.

Lorenz, E. N. (1963) Deterministic nonperiodic flow. *Journal of the Atmospheric Sciences, 20,* 130 - 141.

McFarlane, D. J. (1974) (Ed.) *Motivational Control Systems Analysis.* London: Academic Press.

Nicolis, J. S. (1983) The Role of Chaos in Reliable Information Processing. In Basar, E.,Flohr, H., Haken, H., and Mandell, A. J. *Synergetics of the Brain,* Berlin: Springer-Verlag. pp 330 - 344.

Packard, N. H., Crutchfield, J. P., Farmer, J. D., and Shaw, R. S. (1980) Geometry from a Time Series. *Physical Review Letters, 45,* 712 - 716.

Parker, T. S. and Chua, L. O. (1987) Chaos: A Tutorial for Engineers. *Proceedings of the IEEE, 75,* 982 - 1007.

Peretto, P., and Niez, J. J. (1986) Long Term Memory Storage Capacity of Multiconnected Neural Networks. *Biological Cybernetics, 54,* 53 - 63.

Peretto, P., and Niez, J. J. (1986) Stochastic Dynamics of Neural Networks. *IEEE Transactions on Systems, Man and Cybernetics, SMC-16,* 73 - 83.

Peitgen, H.-O. and Richter, P. H. (1986) *The Beauty of Fractals.* Berlin: Springer-Verlag.

Pimbley, G. H. (1973) On predator-prey equations simulating an immune response. *In Nonlinear Problems in the Physical Sciences and Biology* Lecture Notes in Mathematics No. 322). Berlin: Springer-Verlag.

Schuster, H. G. (1984) *Deterministic Chaos.* Weinheim: Physik-Verlag.

Takens, F. (1981) Detecting Strange Attractors in Turbulence. *In* Rand, D. A., and Young, L. S. (Eds.), *Lecture Notes in Mathematics No. 898.* Berlin: Springer-Verlag, p.336.

Thurstone, L. L. (1927) Psychophysical Analysis. *American Journal of Psychology, 38,* 368 - 389.

Tsuchiya, T. (1984) An exactly solvable difference equation that gives pure chaos for a continuous range of a parameter. *Zeitschrift für Naturforschung, 39a,* 80 -82.

West, B. J. (1985) *An Essay on the Importance of being Nonlinear. Lecture Notes in Biomathematics, No. 62.* Berlin: Springer-Verlag.

2   Choice of a Recursive Core Equation

The self-imposed question that is our starting point in modelling psychophysical data properties is "What is the simplest dynamic structure which might support a diversity of observable input-output relationships, whose parameters are potentially interpretable ?" There is an infinity of solutions to any heuristic question, many of them trivially indistinguishable, but we can narrow down the choice by creating some set of boundary conditions which should be simultaneously satisfied.

Some properties which a single model of sensory intensity might be required to support (without excluding others of special interest in particular modalities or contexts) merely by changing parameter values in the model, and not by altering model structure, are

A1: The psychophysical gain function (often called the psychometric function) is monotone increasing within a bounded range. Psychophysical relations are usually almost (see A4) continuous increasing real functions of physical stimulus inputs, over some range bounded by what are usually termed *thresholds,* or threshold zones.

A2: Response sequences, made in numerical form or quantifiable, to random input sequences, are in varying degrees autoregressive. So-called assimilation and contrast effects have been variously reported.

A3: The linear transfer function, or impulse response function, is typically oscillating about a monotone decreasing mean, and initially positive.

A4: Gain functions can exhibit local breaks (e.g. Hellman and Zwislocki, 1964). When a psychophysical function shows such discontinuities they will in turn be reflected in the sequential dynamics of the system. This can invalidate a linear time series analysis of sequential effects.

A5: To a first approximation, the psychophysical function resembles a segment of a cumulative normal ogive or a logistic curve. Such an approximation has been in use for at least a century, with many second-order corrections.

A6: In some situations, the observer degrades the information content of the input series as it is mapped into the output; there is an upper bound upon what can be transmitted.

A7: The form of the psychophysical function is an overall system gain equation, but closely resembles the sensory receptor transduction equation within it, in some modalities. This means that physiological models of receptor action sometimes are apparently algebraically indistinguishable from psychophysical equations.

A8: The implication of A7 is that there are some post-receptor neural pathways whose overall gain function is almost pure linear and delay.

A9: Any model of sensory intensity should postulate component operations that can, in principle, be realized by a finite neural network. Such operations include summation, multiplication, feedback, integration, and switching.

A10: Accumulating evidence indicates that biological systems with relatively stable outputs are nonlinear, and may exhibit the characteristic features of nonlinear systems, that is, normally functioning in limit cycle and deterministic chaos modes.

A11: Thresholds can show abrupt jumps in long response sequences and sometimes may oscillate.

A12: The subjective Weber function and its associated variance should be derivable from local random perturbations of the system parameters, which input to the real output of the system.

A13: Responses to sinusoid input series may fail to track proportionately in regions of greatest input acceleration.

A14: The traditional psychophysical models (Fechner, Weber, Power laws, Signal Detection, and their derivatives) without fundamental modification, and hence increased structural complexity, can say nothing about A2, A3, A4, A10, A11, A12, A13. They are curvefitting to input-output relations standing outside real time. Models that can meet the criteria listed here without having greater degrees of freedom must start from a different basis.

A15. Clinical evidence, in conditions such as migraine or epileptic auras, indicates that internally-generated transient sensory experiences are a feature of sensory pathways including the frontal projection areas of the cortex, and arise when some features of general regulation are apparently inoperative.

A16: Neural pathways responsible for mediating sensory experience are not fully insulated but are embedded in complex networks into which signals can "leak". There is a background level of activity in such networks through which signals pass. The neural central representation of point sensory peripheral stimuli is extended in space and time.

A17: If a stimulus input series is in a narrow energy band with a bimodal distribution, then a plot analogous to a receiver operating characteristic may be generated without the need to postulate further parameters in the system.

A18: Periodic inputs may reveal hysteresis in outputs which attempt to track accurately the input variations.

A19: If judgments are comparative against an anchor stimulus, and the stimulus range is shifted markedly with respect to the anchor value, then so-called transposition phenomena will occur. In these circumstances the psychophysical function may be positively accelerated for a lower stimulus range, and negatively accelerated for a higher stimulus range.

The objective in constructing a system theory model of sensory intensity, which is in principle valid for any sensory modality, is to meet all the points A1 to A19 from one single process, allowing only minor and realizable parameter changes within the model.

Given these desired properties, which seem to have some generality across modalities and across dimensions within modalities, and in some cases are as much cognitive as sensory, we conjecture that a model which is time dependent is needed, and which to a first-order approximation can generate a cubic gain function. A starting point which is simple in structure and has received repeated consideration in the literature is that considered by May (1974), though its roots are in 19th century models of population dynamics (Verhulst, 1844). This is

$$Y_{j+1} = -a \cdot Y_j \cdot (Y_j - 1), \qquad 0 < Y < 1 \qquad [2.1]$$

where Y is Re, and a is a scalar. This simple form will exhibit complex dynamics, and transitions into bifurcation, cycles, and chaos, as a function of a. In this simple case precise analytic solutions can be obtained for values of a at which phase transitions occur. These are given by May; we do not use them here but note that the qualitative dynamic progression with increasing a in [2.1] will be expected to be analogous in the equation we will employ. There appear to be a limited number of routes to chaos (Ruelle and Takens, 1971. Hao, 1984) but precisely which is taken by a difference equation requires examination, it is not immediately deducible from the form of the equation. After some trial and error, the following form was used.[1] The process

$$Y_{j+1} = -a \cdot (Y_j - 1)(Y_j + ie)(Y_j - ie) \qquad [2.2]$$

---

[1]   We have called it Γ simply because it was the third equation whose

where a is Re, ie is Im, Y is complex (Re,Im), exhibits very complicated dynamics with interplay of the effects of a and e. The boundary conditions are $0 < Y(\text{Re}) < 1$, for simulations here unless otherwise stated $Y_0 = (\zeta, \epsilon)$, $\zeta \leq .5$, $\epsilon < 10^{-8}$, $2 < a < 4$, $0 < e < .5$, and very approximately $ae < 1.7$, in the region of $.5 < e < .7$. Some general properties for very low e, given the constraints ζ and ϵ, are that, for $a < 2$ the value of $Y(\text{Re})$ converges onto a fixed point, for values of $2 < a < 4$ (which are the ones of greatest interest here) it moves variously as a increases onto a limit proportional to a, for a number of recursions $\eta, j = 1, ..., \eta$, which is less than 100, then as a goes beyond 4 onto cycles with some jumps in periodicity by bifurcation (doubling), and then, when a is very near to 6, (for $e = 0$) into chaos. For some values of ae it explodes and goes rapidly to astronomical values which could not be biologically meaningful. The starting value ϵ is not critical for all recursions examined in the stable (a, e) region, provided that ϵ is very small. Simulation of [2.2] needs care because ϵ can be near to the level of rounding-off errors in single precision software; for this reason we can get slightly different results on different computers.

The reasons for guessing at this form are simultaneously

(i) it has minimum complexity consistent with producing a cubic-like gain function of $a \Rightarrow Y_\eta$, where η is the fixed number of iterations of [2.2] being studied,[2]

(ii) the presence of at least one complex conjugate pair, as suggested from transfer function analyses in psychophysics (Gregson, 1984, Gregson and Gates 1985)

(iii) tractability in computer simulation, requiring only exploration of the parameter space $\{a, e, \eta\}$

(iv) potential separability of signal and noise network components by mapping onto the Re and Im parts of Y_η

---

properties we explored, and is of order 3 in Y. The name has nothing to do with Γ functions in distribution theory.

[2] The form $Y_{j+1} = a \cdot Y_j^2 (1 - Y_j)$ is examined by Marotto (1982) because it exhibits threshold phenomena, and has other properties of interest in population dynamics. He notes that for $a > 5.89$ the process goes into chaos, but beyond $a = 6.54$ the chaos is nowhere dense. This finding led Marotto to suggest that there is more than one sort of chaos. It is possible to treat [2.2] as a special case of

$$Y_{j+1} = a \cdot Y_j^m (1 - Y_j)^n \qquad [2.2a]$$

where $n, m > 1$. which if Y is Real, shows thresholds and chaos and then nowhere dense chaos with increasing a. [2.2a] has not been studied in the case where Y is complex as in [2.2].

(v) retention of the desired dynamic properties of recursive difference equations like [2.1]

(vi) compatibility with some findings in neurophysiology, for example by Freeman (1975).

Some of the properties of [2.2], given its initial conditions, should be noted. Y(Im) has much lower (10^{-8}) relative magnitudes at the limits to which it tends, as compared with Y(Re). Y(Re) can run to a stable point limit when at the same time, with increasing j, Y(Im) goes into cycles; even more complicated patterns in which the dynamics of the Re and Im components are qualitatively different can be found with suitable values of a and e. Y(Im) can show intermittency patterns, a phenomenon in which runs at one value of Y(Im) are interspersed with periods of oscillating; this pattern can be associated with the presence of a strange attractor. The phenomenon is of potential interest because it is known to arise in neural networks.[3] The limits to which Y(Re,Im) runs with increasing j are independent of the starting point Y_0, but the recursion J at which stability is reached will depend to some extent on Y_0. Hence the choice of η, the value of j at which the recursion is stopped, may be before or after J. For simulations in Table 2.1 η has been set at 50, which is usually well beyond J.

$\{Y(\text{Re,Im})\}$ is a complex time series, and after J is representable as an autoregression in Y plus *noise*. Such a representation is not generally necessary to reveal structure, as from inspection of serial plots like Figures 2.1 to 2.4 or phase space diagrams the periodic or quasi-periodic behaviour of [2.1] or [2.2] can readily be seen. When the process goes into determin-

---

[3] Cowan (1968) in considering the statistical mechanisms of the activity of large neural assemblies advanced an 'equation of motion' which in slightly altered notation is

$$Y_{r,j+1} = Y_{r,j} + (\gamma_r + \beta_r^{-1} \sum_{s}^{N} \alpha_{sr} Y_{s,j}) Y_{s,j} (1 - Y_{s,j}) \qquad [2.2b]$$

where the recursion time is a mean intercellular transit time for neural activity, γ and β are growth coefficients, α_{sr} is a coupling coefficient between the neurons s and r, and input to r enters nonlinearly through γ_r. Y_r means the sensitivity of neuron r in a net, so [2.2b] has to be integrated over a net of N elements to resemble in its purpose the form [2.2] used in this monograph. The interesting parallel is in the cubic terms in Y. There are a number of constraints imposed on [2.2b] to make it a viable model which are discussed by Cowan; the system based on [2.2b] can also be thought of as a system of coupled oscillators.

Table 2.1   (simulations on PUP 11/34, single precision) Y_{j+1} behaviour: $Y_1=(.5,\epsilon)$, n=50,

a	2.3	2.7	3.1
0	converges to .565 in 14 trials	converges to .629 with weak period 2 in 36 trials	period 2 (Re) .77,.54 immediately
.1	converges to $(.024,-4.7\times10^{-10})$ after 14 trials	converges to $(.028,4.7\times10^{-10})$ after 12 trials	converges to $(.033,0)$ after oscillating for 13 trials
.2	converges to $(.105,-3.7\times10^{-9})$ after oscillating in (Im) for 16 trials	converges to $(.137,0)$ after oscillating in (Im) for 26 trials	oscillating at $(.194,0<\text{Im}<1.49\times10^{-8})$ irregularly
.3	converges to .277 in (Re) with period 2 in (Im); $(1.49,2.89)\times10^{-8}$	converges to $(.413,0)$ after oscillating in (Im) for 27 trials	converges to $(.546,-1.49\times10^{-8})$ after oscillating in (Im) for 11 trials
.4	converges to .462 in (Re) with period 3 in (Im); $0<i\omega<-1.49\times10^{-8}$	converges to .563 in (Re) with period 2 in (Im); $\pm2.98\times10^{-8}$ after 4 trials	converges to $(.637,-2.98\times10^{-8})$ after 8 trials
.5	converges to $(.569,0)$ after oscillating in (Im) for 10 trials	converges to $(.641,0)$ after oscillating in (Re,Im) for 25 trials	oscillates at period 2 in (Re) around .69, and at period 3 in (Im) $(-2.98<i\omega<2.98)\times10^{-8}$
.6	converges to .638 in (Re) and irregular aperiodic oscillations in (Im) $(-2.89<i\omega<2.98)\times10^{-8}$	period 2 in (Re) $(.83,.49)$ and period 4 in (Im) $((-2.9,7.4,-5.9,11.9)\times10^{-8})$ after 34 trials	explodes in (Re,Im).

3.5	3.9	comments
period 4 (Re) .50,.87,.38,.82 immediately	limit cycle in reals.	Y_{j+1} $=-a(Y_j)(Y_j-1)$
converges to $(.038,9.3\times10^{-10})$ after oscillating for 22 trials	converges to $(.544,0)$ after oscillating for 42 trials	Y_{j+1} $=-a(Y_j+(0,i\omega))$ * $(Y_j-(0,i\omega))$ * (Y_j-1)
converges to $(.524,-1.4\times10^{-8})$ after oscillating in (Im) for 30 trials	converges to $(.631,1.49\times10^{-8})$ after oscillating in (Im) for 10 trials	as above
converges to $(.630,-1.49\times10^{-8})$ after oscillating in (Im) for 9 trials	converges to $(.686,0)$ after oscillating in (Im) for 21 trials	as above
slow convergence in (Re) around .690, oscillating irregularly in (Im) at $(-5.96<i\omega<5.96)\times10^{-8}$	period 2 in both (Re,Im); $\begin{cases}.60,5.96\times10^{-8}\\.81,-2.98\times10^{-8}\end{cases}$	as above
period 2 in both (Re,Im); $(.87,0),((.45,-1.49)\times10^{-8})$ after oscillating for 26 trials	period 4 in (Re); $(.90,.40,.96,.18)$, irregular (possible limit cycle) in (Im).	as above
explodes in (Re), converges to 0 in (Im)	explodes in (Re,Im)	as above ae< 1.7 apparently needed

istic chaos, however, for high a values, then distinguishing complicated but strictly periodic patterns (after a number of bifurcations) from chaos can be facilitated by time series analysis, and by computing $\bar{\lambda}$ (See [2.5]). The intermittency phenomenon can be represented by a distribution of dwell times in runs of either stationary or oscillating or chaotic behaviour.

The Form of the Y Response Surface for $\{a, e \mid \eta\}$

If [2.2] runs to a limit of η recursions then the result is $Y_{obs.\eta}$ which as defined is complex. It may be single-valued or oscillating in one or both components. For a fixed η, over the parameter space $\{a, e\}$ we may thus construct a bivariate response surface $\mathbf{P}_{Y,\eta}$ of $\{Y_{obs.\eta} \mid a, e\}$, and this at any point will have four values; $minY(\text{Re})$, $maxY(\text{Re})$, $minY(\text{Im})$, $maxY(\text{Im})$. There are thus in theory four distinct response surfaces, but they are not distinguishable at all (a, e) points, and not all of equal interest. Figures 2.5 and 2.6 depict local estimates of $Y(\text{Re})$ for $\eta = 1000$ and $\eta = 1001$ which pictorially captures some of the fluctuations.

Sections γ (constant $a = \gamma_a$, constant $e = \gamma_e$), through $\mathbf{P}_{Y,\eta}$ are important, as rows or columns of the (a, e) grid as in Table 2.1 constitute special restricted models to which it may be desired to give interpretation. We will examine the possibility of generating some psychophysical gain functions from sections γ_a or γ_e in more detail in later chapters.

There is relatively little value in exploring $\eta > 50$ because most of the Y series reach a stable limit or cyclic pattern well before $\eta = 50$. Very small values of η can show more disparities between the $minY(\text{Re})$ and $maxY(\text{Re})$ values, as is inferrable from Figures 2.1 to 2.4.

The properties of one-dimensional non-invertible maps of the interval have been successfully used in the study of deterministic chaos, including chaos in systems of higher dimensionality such as the Lorenz (1963) equations. Shaw (1981), and more recently Fischer and Smith (1985), have shown in detail that one-dimensional maps exhibit much of the complicated dynamic behaviour observed in systems of higher dimensionality. Extensive general reviews of the entropy production of such systems are presented by Schuster (1984), Hao (1984) and Mayer-Kress (1986).

Examples of chaotic behaviour in one-dimensional mapping functions have been collected by May and Oster (1976), and Metropolis, Stein and Stein (1973). Amongst others May (1976), Singher (1978), Fiegenbaum (1978, 1979), Shaw (1981), Ott (1981) and Grassberger and Procaccia (1983) have identified many fundamental properties of [2.1], including the identification and calculation of fixed points, critical points, Lyapunov coefficients and Kolmogorov-entropy. More recently, extension of the analyses to consider the case where noise is also injected randomly into the process

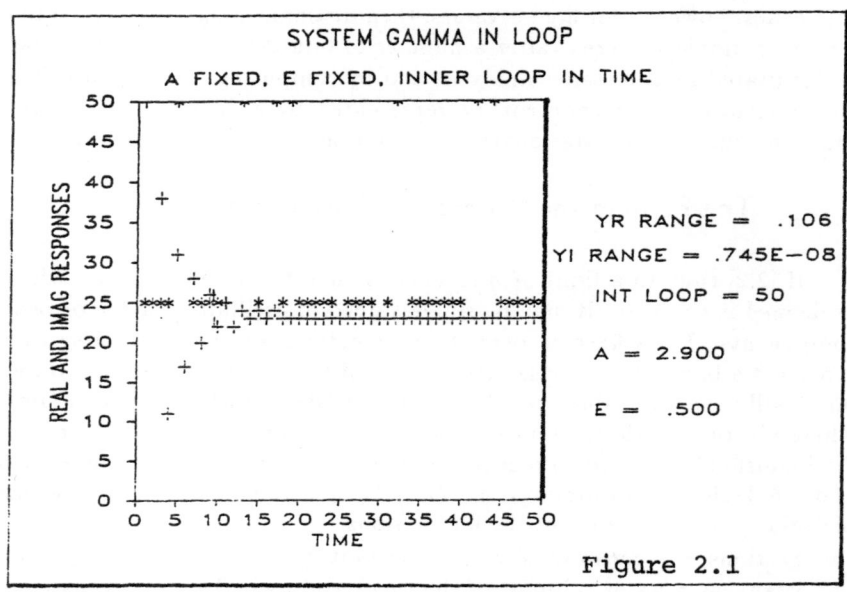

Figure 2.1

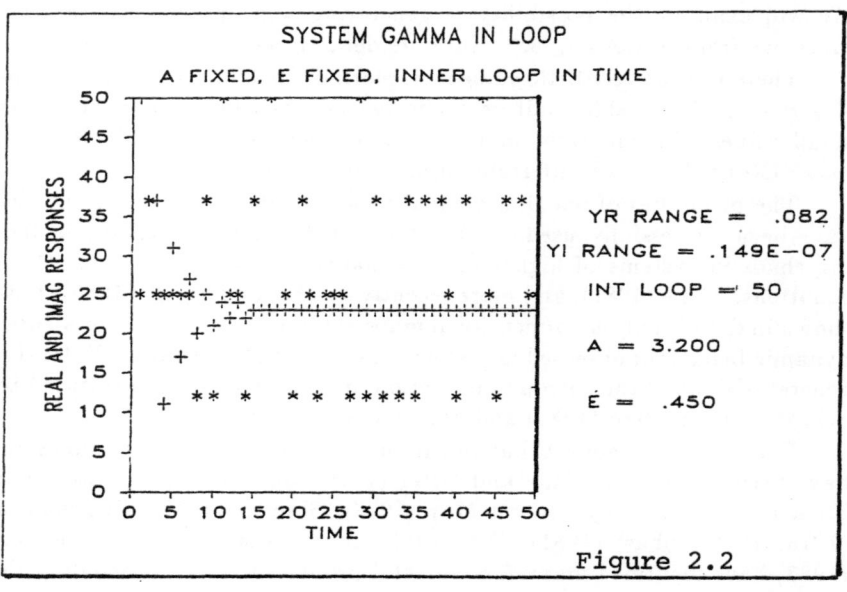

Figure 2.2

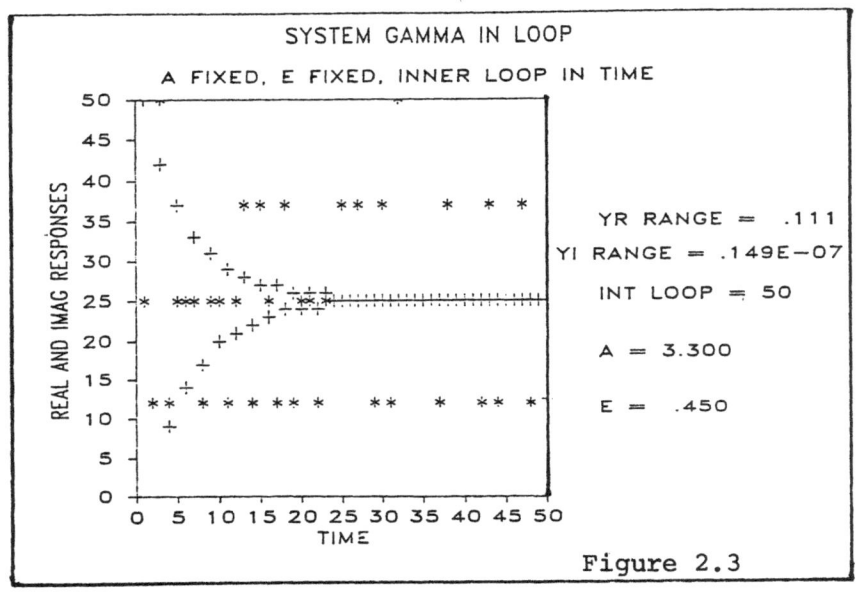

Figure 2.3

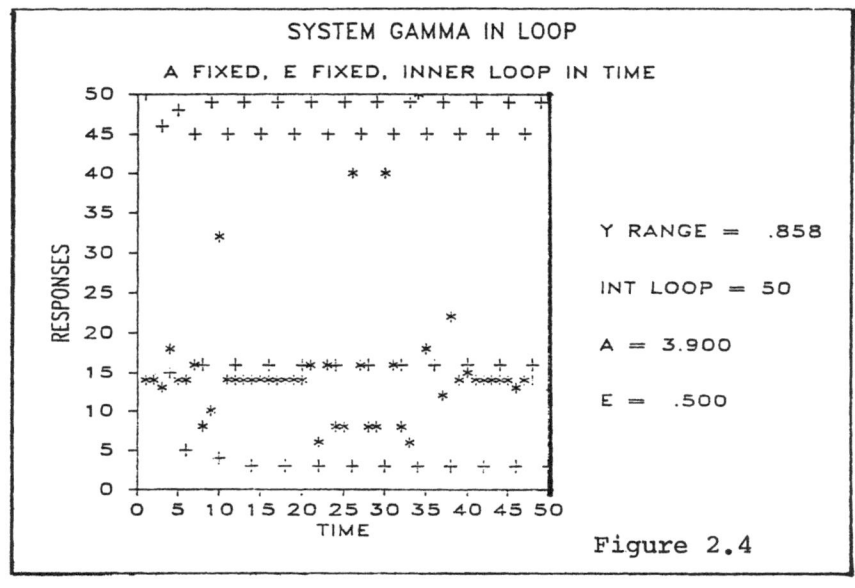

Figure 2.4

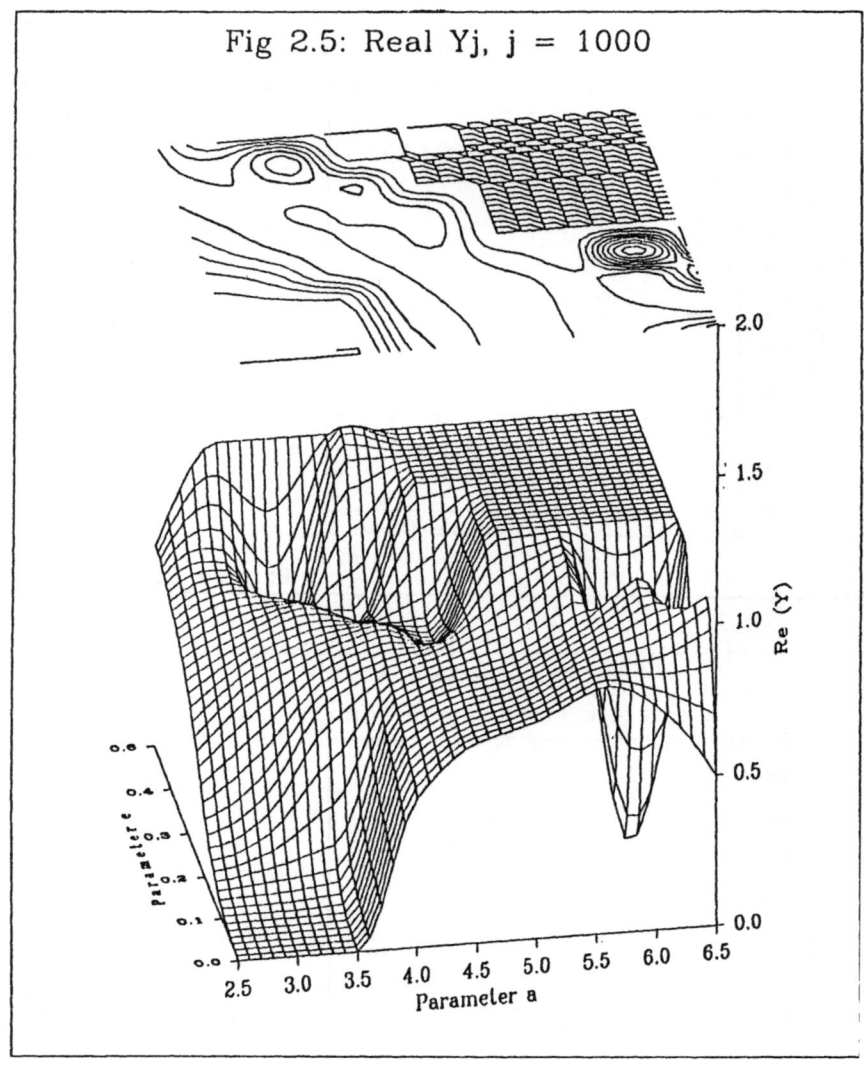

Fig 2.5: Real Yj, j = 1000

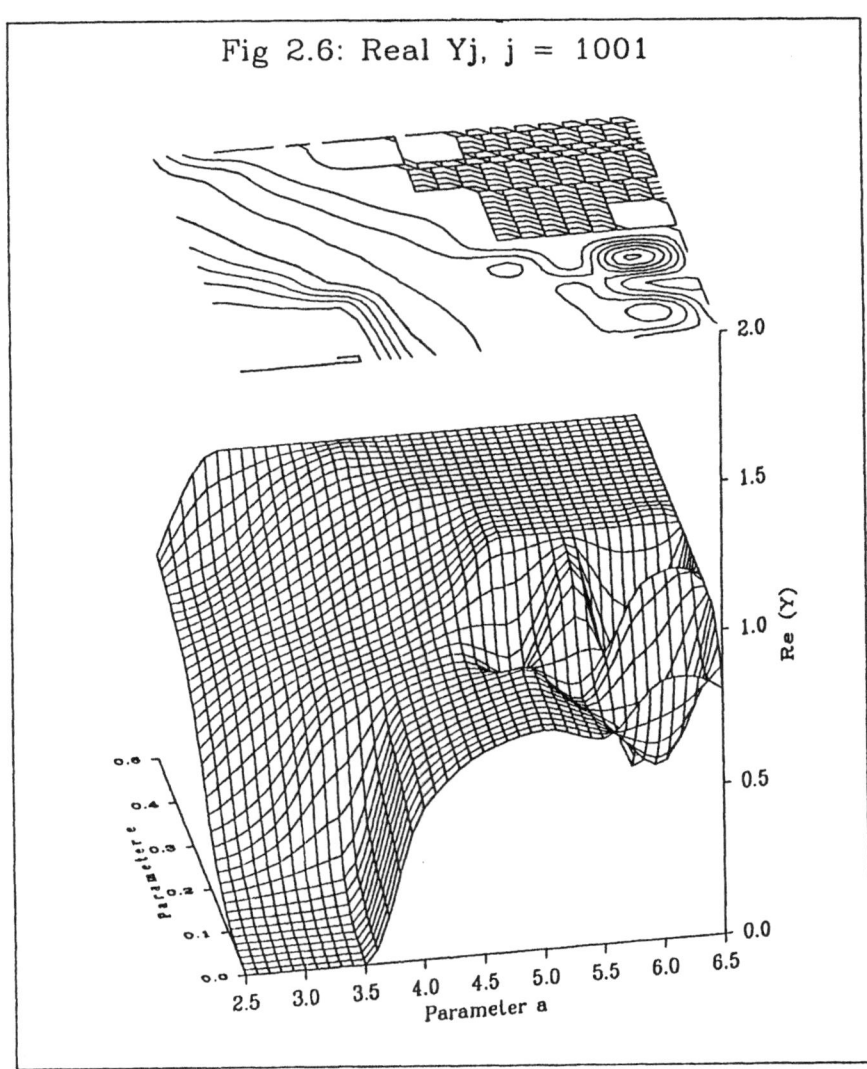

Fig 2.6: Real Yj, j = 1001

(Argoul and Arneodo, 1986) is of particular neurobiological interest.

The motivation for this activity is obviously not purely mathematical, it stems from a closer look at a rich diversity of phenomena which occur in dissipative systems observed in both the physical and the biological sciences, so this search is both for novel and subtle mathematical results and for the dynamics of real-world systems that might correspond and were previously thought intractable as objects suitable to be modelled. As Lopes (*J. Math. Psychol.*, 1984, *28*, 483) has remarked,

> "What makes for good science anyway ?.... one factor seems uniform in all the successful subfields within psychology. That is an intense and respectful interest in the phenomena that the science seeks to explain".

May (1976), May and Oster (1976), Zeeman (1976), Freeman (1979, 1983), Mackey and Glass (1977), and Babloyantz (1986) are examples which are in the spirit of this study.

For example, May and Oster (1976) used [2.1] as a model for biological population dynamics where populations have discrete non-overlapping generations. Mackey and Glass (1977) associated the onset of disease in physiological systems with bifurcation points in the dynamics of first-order differential delay equations, and Zeeman (1976) considered the qualitative analogies between the nonlinear Duffing and Van der Pol oscillators and some psychophysiological processes such as memory recall, mood, manic depression and even anorexia nervosa. In this latter case Zeeman was partly motivated by a connection into his earlier work on catastrophes in the same syndrome (collected in Zeeman, 1977).

Complex variables and parameters have been studied in an extension of the Lorenz (1963) equations by Fowler *et al* (1982) and it was there found that behaviour in the imaginary bifurcation sequence arose which was significantly different from that found in the reals. Mandelbrot (1980) had reported results for the logistic equation [2.1], but with both a and X complex. Schwartz (1984, and earlier work cited therein) has proposed a complex logarithmic model for primate visual representation through a mapping of 2-dimensional space onto 2-dimensional space in the striate cortex. There are thus physiological and mathematical precedents for exploring complex functions like [2.2].

The investigation of the dynamics of [2.2] was prompted by the possibility that it might serve as the core of a system (see Chapter 3), involving the action of a feedforward network with one loop which could generate psychophysical transfer functions mediating sensory intensity, if input (stimuli) is mapped $1 \Leftrightarrow 1$ onto a and the $Y(Re)$ is mapped $1 \Leftrightarrow 1$

onto output (responses) [4] .

The parameter ϵ in [2.2] may be thought of as an error term, inducing oscillations at a secondary level, which are observed (Werner and Mountcastle, 1963) as being intrinsic to neural network activity, but the interpretation is a heuristic and not strictly necessary. We may treat e as having a definable stochastic distribution in time, or as a fixed parameter. For the results presented in this chapter, e has been treated as a constant for the full evolution of a particular map. Later the consequences of a variable e are examined. The a parameter is a scalar gain or destabilizing parameter as is usual in recursive mappings (May and Oster, 1976, Hao, 1984, Chapter 4).

Since [2.2] is cubic in $Y_j(\text{Re},\text{Im})$, an analysis similar to that given by May (1976) on the logistic function [2.1], quickly reveals [2.2] to be highly intractable analytically. It is useful to establish computationally, in a qualitative sense, some of its major properties and thus to facilitate its comparison with other functions of potential interest, as relatively little is yet known about functions like [2.2].

For simplicity, considering [2.2] with the $Y_j(\text{Re},\text{Im})$ replaced by the corresponding real parts x_j $(= Y_j(\text{Re}))$ only, will clarify matters. That is,

$$x_{j+1} = a \cdot (x_j - ie)(x_j + ie)(1 - x_j) \qquad [2.3]$$

Basic mathematical properties of [2.3]

The mapping function

Singher (1978) reintroduced the use of the Schwarzian derivative in the context of the analysis of one-dimensional maps on the interval. The derivative is defined as

$$S_f = \frac{f'''}{f'} - \frac{3}{2}\left(\frac{f''}{f'}\right)^2 \qquad [2.4]$$

and if S_f is negative it is considered to show good "expansive" properties (Guckenheimer, 1980). Many mapping functions, including [2.1], have a negative S_f for all x.

---

[4] Some $a \mapsto Y$ mappings for fixed e resemble psychometric functions in the range $2 < a < 4$. Such functions may be regarded as section planes through the surface of part of Figures 2.5 and 2.6, which show, *inter alia*, that oscillations continue for large Y, given high a and only low e.

In the case of [2.2] the use of [2.4] is invalidated because the variable
is complex. However, considering only the reals as in [2.3], the derivative
is everywhere negative only for the range

$$e^2 < \frac{1}{3}$$

or

$$|\, e\, | < 0.5774 \,.$$

For this reason, consideration is restricted to [2.2] in the range $0 < e < .6$,
it being noted that the function is symmetrical about $e = 0$.

When $e = 0$, [2.3] becomes a single-peaked cubic mapping function
with $f(0) = f(1) = 0$, which is similar to [2.1] although not symmetrical
over the interval. As e is increased above 0 in [2.3], $f(1)$ remains 0, and
$f(0)$ is always positive.

For $0 < e < \frac{1}{\sqrt{3}}$ a local maximum and minimum occur for the mapping
within the unit interval. At $e = \frac{1}{\sqrt{3}}$ the maximum and minimum converge
to an inflection point at $x = \frac{1}{3}$. That is, the mode in the mapping function
has then vanished.

As a is increased, the mode becomes more and more peaked (imagine
a cross-section through $e = 0$). By analogy with the behaviour of [2.1],
it would be expected that trajectories would exhibit period doubling and
eventually chaos as the hump is accentuated.

For [2.3] the maximum value for a is 6.75, given $e = 0$, but for the
complex-valued function [2.2] the maximum is approximately 6.54. Introducing $e > 0$ apparently modifies the region which Marotto (1972) called
'nowhere dense chaos'. The corresponding imaginaries are of the order of
10^{-9}, provided that ϵ is made to be small. However because of the introduction of e (and hence the consequent release of the constraint that $f(0) = 0$),
at any a value except $a = 0$, a corresponding value for e can be found which
causes the system to increase rapidly without limit; in figurative language
it 'explodes'.

Figure 2.7 shows the boundary set of admissible values of a, e, which we
have called $\{a, e\}_B$, for both [2.2] and for [2.3] [5] All values of $\{a, e\}_B$ were
found by simulation to an accuracy of 0.01, such that the absolute value
of the output $|\, Y_{j+1}\, |$ or $|\, x_{j+1}\, |$ was ≤ 1.00 after 1000 iterations. This
boundary depends on setting ϵ small in Y_0; if $\epsilon > 10^{-3}$ then simulations
suggest that the process can run to explosion at some a, ϵ values which are

---

[5] Most of the computer graphics in this chapter, and some quantitative
results, were derived by Ian Price, to whom warm thanks are due, and are
also reported in a slightly different form in Price and Gregson (1988).

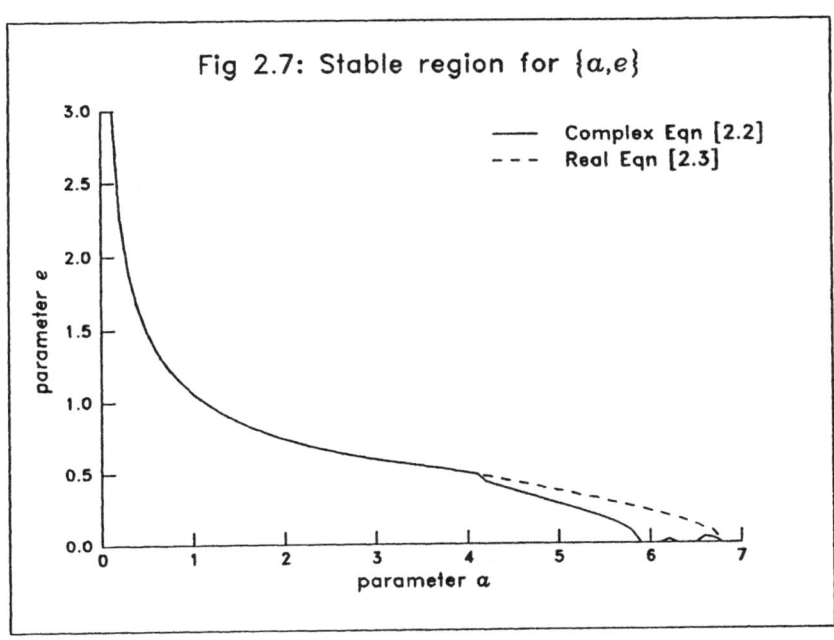

Fig 2.7: Stable region for $\{a,e\}$

in the stable region shown in Figure 2.7 . Note that the boundary values are the same for both [2.2] and [2.3] up to $\{a, e\} = \{4.0, 0.5\}$, beyond which point the imaginary components exercise a greater influence.

Given the constraint that $\mid e \mid < \frac{1}{\sqrt{3}}$, the $\{a,e\}_B$ values lie within the bounds

$$\begin{cases} 2.958 \leq a \leq 6.54 \\ 0.6 \geq e \geq 0.0 \end{cases}$$

where e decreases as a increases.

Figures 2.5 and 2.6 depict the real output of [2.2] after 1000 iterations and after 1001 iterations respectively (where $\hat{Y}_{j+1}(\text{Re}) = 1.00$ if $Y_{j+1}(\text{Re}) \geq 1.00$). These two values of j are chosen to show the qualitative fluctuations between local minima and maxima within which the system is oscillating. Many points in the a, e parameter space imply a convergence onto stable limits (absorption onto a point), however the figures also show quite dramatically the form of the bifurcations occurring at a, ϵ values in the neighbourhood of the $\{a, \epsilon\}_B$ locus.

Note that in Figures 2.5 and 2.6 the upper part of the diagram is a contour map projection of the 3-dimensional surface below. The pocket within $5.92 < a < 6.04$, $.234 > \epsilon > .225$ appears to be a region of local stability beyond $\{a, \epsilon\}_B$. This particular local stability may more critically

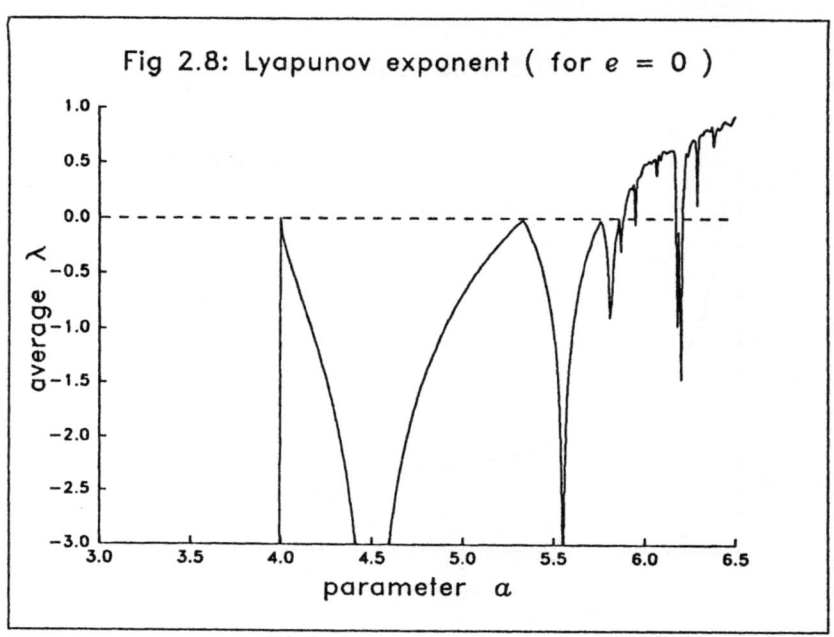

Fig 2.8: Lyapunov exponent (for $e = 0$)

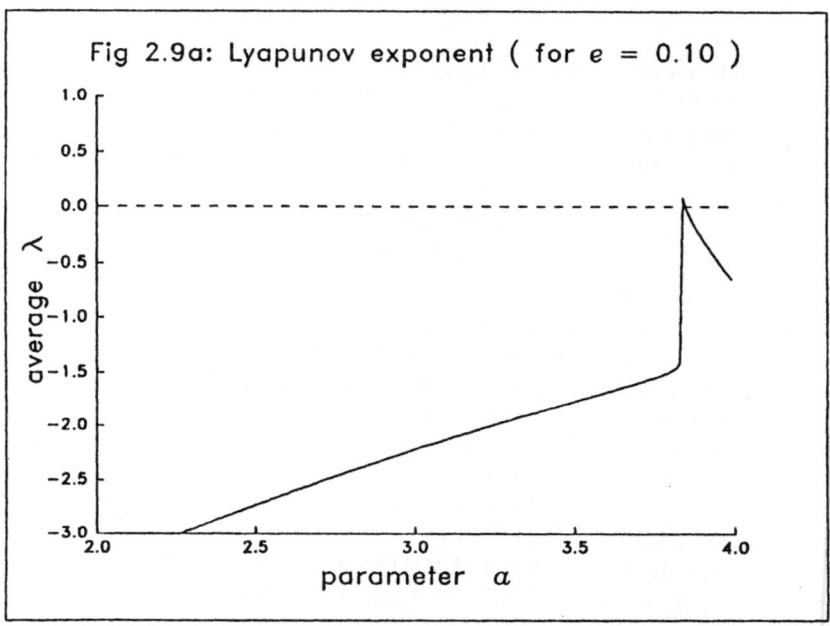

Fig 2.9a: Lyapunov exponent (for $e = 0.10$)

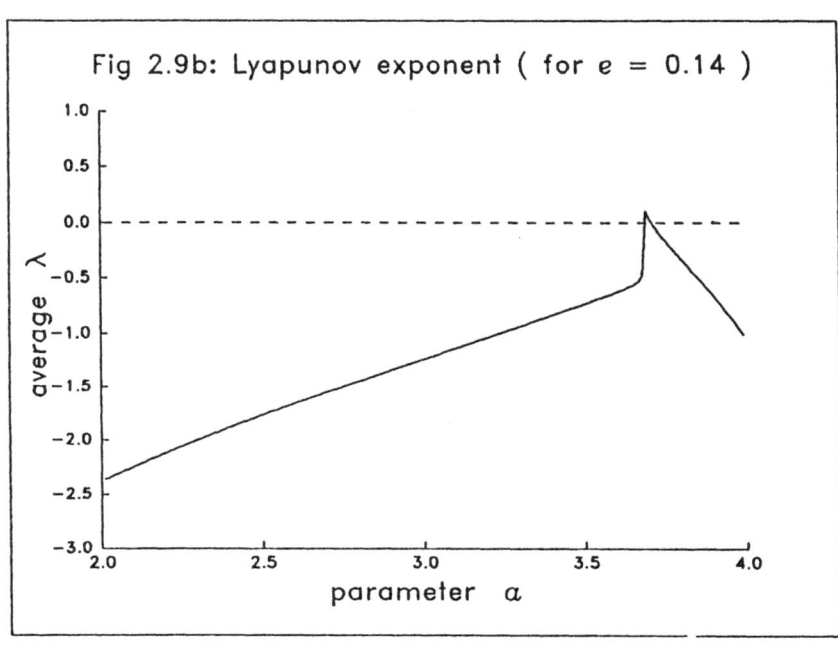

Fig 2.9b: Lyapunov exponent (for $e = 0.14$)

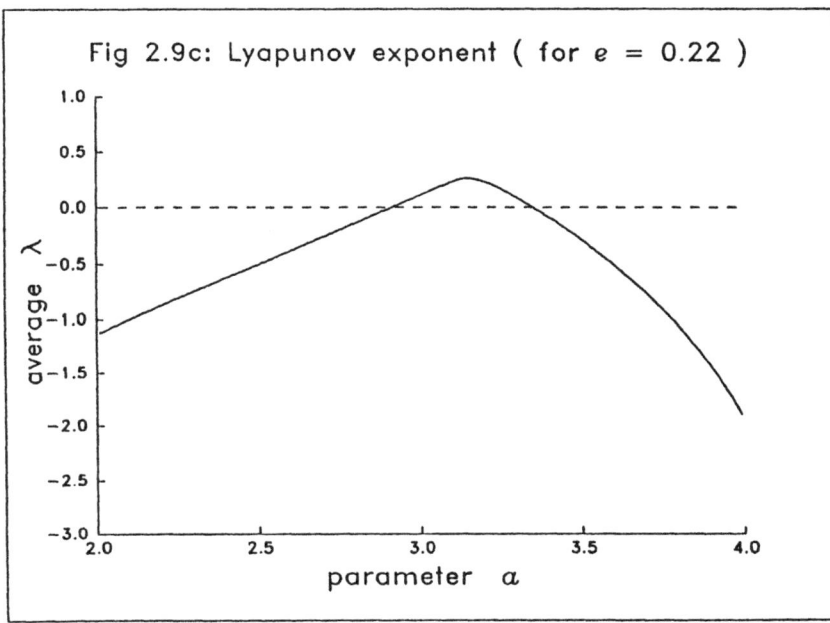

Fig 2.9c: Lyapunov exponent (for $e = 0.22$)

depend on ϵ being very small in Y_0.

The Lyapunov Index

It is required to know if a system is in a cyclic or in a chaotic (strange attractor) phase. The question is in principle answered by measuring both the information at the starting point of a definable dynamic sequence, and the loss of information at each iteration of the process[6] The appropriate mathematics stems from Russian work, and the type of index we seek to calculate is called a Lyapunov coefficient or exponent. The relevant literature is now voluminous; for reviews see Schuster (1984), Hao (1983), Arnold and Wihstutz (1986), and Kreuzer (1987). [7] This coefficient $\bar{\lambda}$ is a measure of exponential growth rates. When $\bar{\lambda}$ is negative the system converges to a stable attractor, a point or a limit cycle, when $\bar{\lambda}$ approaches zero the system has less and less competence to absorb transient perturbations, and for $\bar{\lambda} \geq 0$ the initial uncertainty persists indefinitely. Positive $\bar{\lambda}$ implies , therefore, that a system is self-aggravating in its capacity to destroy information about its own previous states. A definition of $\bar{\lambda}$ in continuous one-dimensional states is

$$\bar{\lambda} = \lim_{n \to \infty} \frac{1}{n} \sum_1^n \left(\frac{log_2 \mid dy/dx \mid}{T_j} \right), \qquad y = F(x) \qquad [2.5]$$

but for systems which are in a bounded space, such as a lattice, other expressions which have the same properties in terms of local changes in entropy have to be, and can be, derived (Fogelman-Soulié, 1986). There is no general expression for $\bar{\lambda}$.

Shaw (1978) introduced the parameter $\bar{\lambda}$ as a measure of the average information change after each iteration of the interval map, and noted that in the one-dimensional case it was equivalent to an index previously derived by Oseledec (1968) with origins in the work of Lyapunov in the 19th century. The information change per iteration is here computed from the limit of the mean log_2 of the gradient of the map and may be defined for computational purposes as

$$\bar{\lambda}_\Gamma = \lim_{n \to \infty} \frac{1}{n} \sum_i^n log_2 \left| \frac{\Delta^1 Y_{j+1,j}}{\Delta^1 Y_{j,j-1}} \right| \qquad [2.6]$$

---

[6] The time for one iteration, T_j, obviously is a factor in estimating the rate of information loss in real time, but can without loss of generality be set at 1.

[7] Kreuzer has also developed a method which rests on Markov chain analyses of the system trajectory in a discretized phase space. This may be more readily computable on small data samples.

For $\epsilon < 10^{-6}$, and over a range of a, $\bar{\lambda}$ showns interesting behaviour with some local regions of stability outside the main range $0 < a < 4$. This is shown in Figure 2.8. Computed $\bar{\lambda}_\Gamma$ for a range of non-zero ϵ values are shown in Figures 2.9a, 2.9b, 2.9c. Each graph is composed of 350 points spaced 0.01 apart, with 1000 iterations of the complex map [2.2], generated on a DEC20 with single precision, for each point. Negative values of $\bar{\lambda}$ indicate stable periodic orbits in the phase space, and the degree of negativity is a monotone measure of the stability of the periodic orbit in the face of small extraneous perturbations; from [2.5] we can derive the expected number of cycles before initial information is completely lost.

Bifurcation of periodic orbits (limit cycles) are marked by points of tangency with $\bar{\lambda} = 0$, that is, an average slope of unity. Positive $\bar{\lambda}$ values are associated with chaotic régimes (a, ϵ values) in which information is being produced, and the information contained in the initial state at Y_0 is lost. The point $Y_0 = (0.5, \epsilon)$ is a stable fixed point for the parameter combination $\{a, e\} = \{4.5, 0.0\}$ – the first sink in the graph of Figure 2.8. As e is increased the locus of the fixed point decreases. Note that the critical parameter value a_c for the accumulation of period 2^n cycles (i.e. where $\bar{\lambda}$ crosses zero) is dependent upon the parameter e. At $e = 0$, a_c is approximately 5.89. Beyond this value of a trajectories may be considered chaotic.

As e is increased, points in Figure 2.9a which were initially tangent to $\bar{\lambda} = 0$, that is, bifurcation points, may be greater or smaller than zero in Figures 2.9a, 2.9b and 2.9c. This suggests that the parameter e is capable of inducing chaotic behaviour at relatively low a values. This chaotic behaviour occurs in the imaginaries at $\{a, e\} = \{3.31, 0.20\}$. So, at the parameter values corresponding to the first positive peak in Figure 2.8, the reals quite rapidly approach a stable limit at .33741, whereas the imaginaries exhibit somewhat bizarre behaviour. The probability of a biological system landing on this precise point is very low.

Figures 2.10 and 2.11 give a $\bar{\lambda}$ surface for a range of a, e values in the area of major psychophysical interest; the same surface is viewed from two directions.

Real and Imaginary Trajectories

Figures 2.12 to 2.15 give examples of the trajectories in both the real and imaginary components of [2.2]. The points in the a, e parameter space were chosen from regions with $\bar{\lambda} > 0$ (see Figure 2.8 for comparison), and include the first 100 points of the iterated map. The examples show the increased activity in the phase space of each of the components as a is increased and e is in the neighbourhood of an explosive value. We draw attention to the expansion and topology of the trapping region in

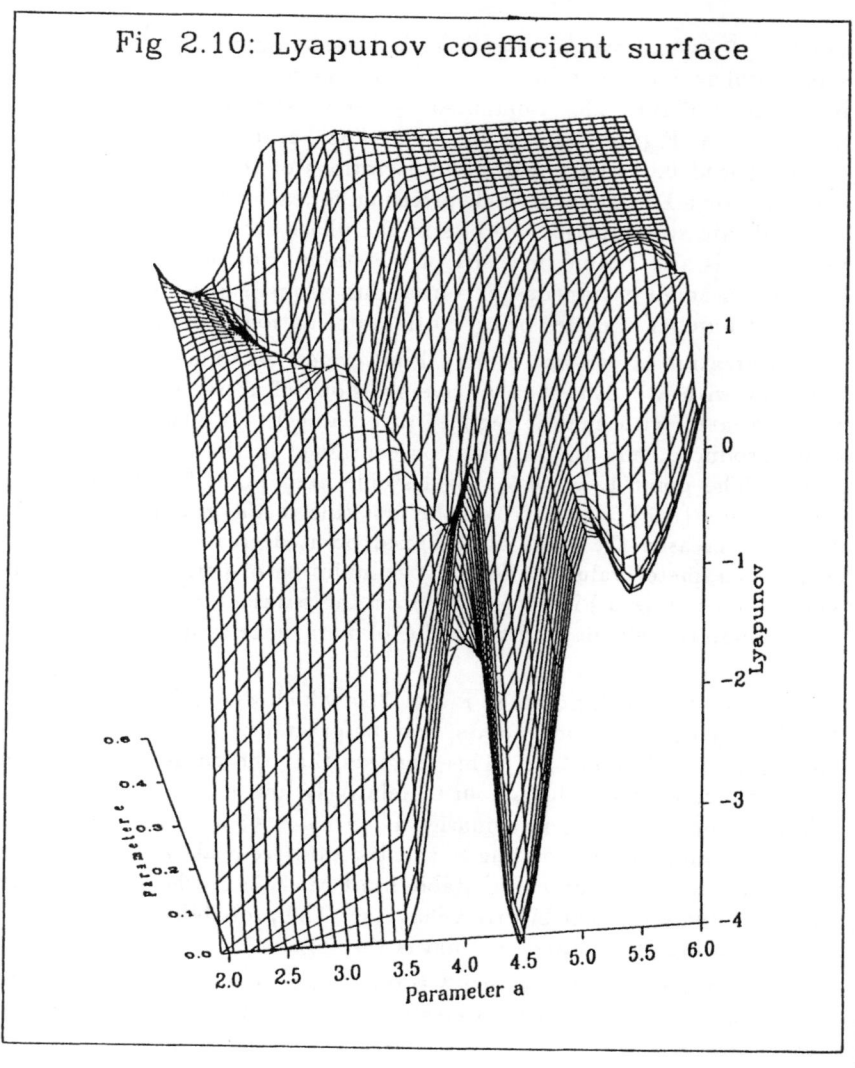

Fig 2.10: Lyapunov coefficient surface

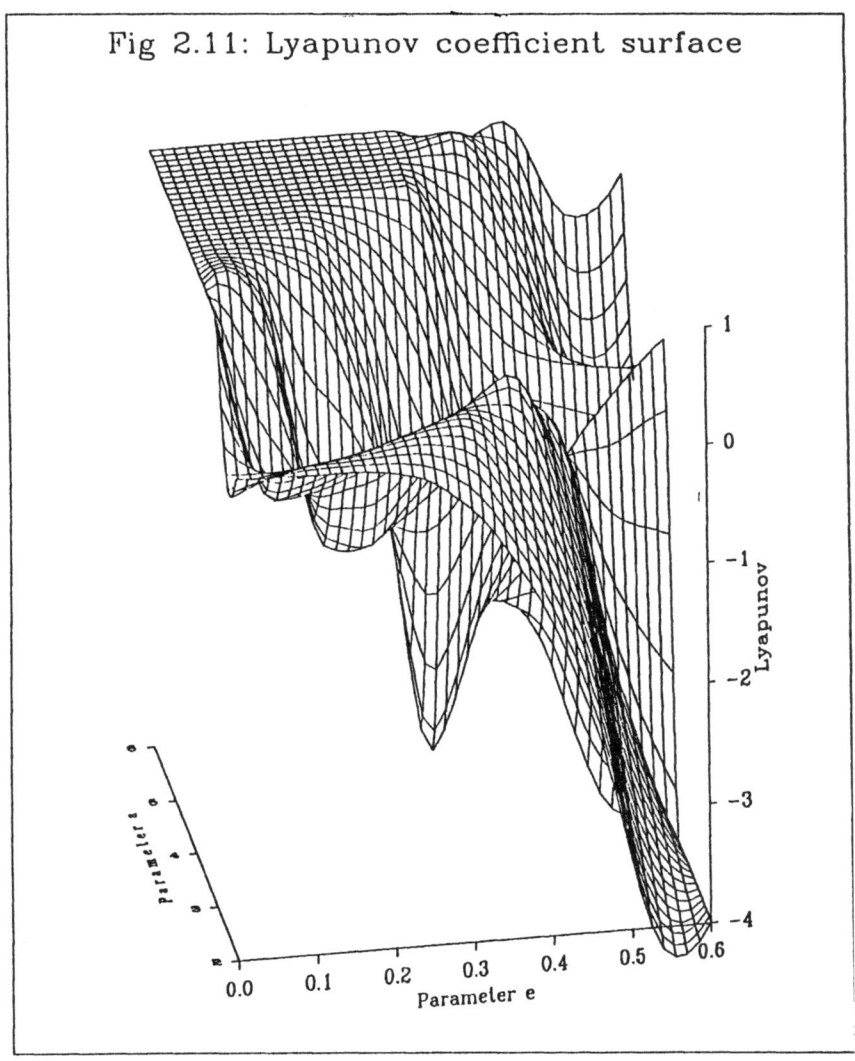

Fig 2.11: Lyapunov coefficient surface

the reals as a is increased, and the appearance of bands of activity. This region eventually spreads to cover, or 'shadow', the interval (over at least $.2 < Y < 1.0$ at $(a, e) = (6.543, 0.0)$.) Compare results by Argoul and Arneodo, 1986. From the perspective of the experimental psychologist we do not need to know if it eventually covers the whole unit interval, nor do we need to know the location of critical points to a resolution finer than two decimal places.

Figures 2.12 to 2.15 also show that the behaviour of the imaginary components may exhibit quite high periodicity or even aperiodic behaviour at parameter combinations for which the reals are quite well behaved. For instance at $(a, e) = (2.958, 0.6)$, as in Figure 2.15, the reals are oscillating on a P_2 basis between 0.19017 and 0.949 whereas the imaginaries show a considerable diversity of behaviour (see also Figure 2.11).

The behaviour of the imaginaries does not show any obvious pattern of development that might be considered complementary to that of the reals; there is no apparent period-doubling in one part which matches that in the other. As previously observed, the imaginary components induced here are of very small relative magnitude (of the order of 10^{-9}). If they are small the process is self-stabilizing in $Y(\text{Re}, \text{Im})$ for a wide range of (a, e) but there is a point at which if $Y(\text{Im}) > \zeta$, then the process is explosive, and $\zeta = f(a, e)$. A model of real -world phenomena, in massive neural networks serving as a substrate for psychophysics, might require that we postulate some pure amplification of this component on the output side of a system model. Figure 2.16 gives an example of the same trajectories as the corresponding Figure 2.12 but viewed in the complex plane.

One interesting parameter combination is $(a, e) = (4.0, 0.5)$. This point yields an immediate indefinitely persisting oscillation in the reals between zero and one, while the imaginaries remain at zero. That is, the complex mapping function at this point exhibits a purely real, binary behaviour sequence.

The estimation of the dimensionality of an attractor in the phase space is obtained by plotting it in an embedding space of m dimensions, defining these by the variables $(Y_j, Y_{j+1}, Y_{j+2}, ..., Y_{j+m-1})$, $j = 1, ..., M$, and then filling this space of volume 1^m (Y^m given that $0 < Y(\text{Re}) < 1$) entirely with a gridwork of contiguous little boxes, each of side r. Every point to which the trajectory goes then lies in a hypercube, and is scored as such. In the limit as the length of a cube side becomes infinitesimally small, then approximately

$$Q(r) \sim r^D \ , \qquad\qquad [2.7]$$

where

$$Q(r) = \lim_{n \to \infty} \cdot r^m \cdot \sum_{i=1}^{r^{-m}} N(Y_i, r) \qquad\qquad [2.8]$$

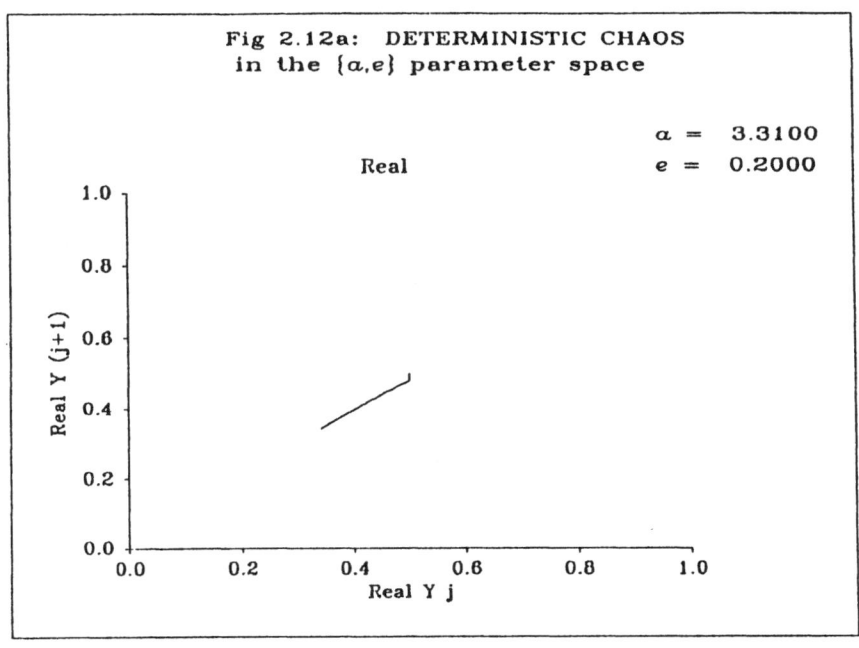

Fig 2.12a: DETERMINISTIC CHAOS
in the {a,e} parameter space

a =   3.3100
e =   0.2000

Real

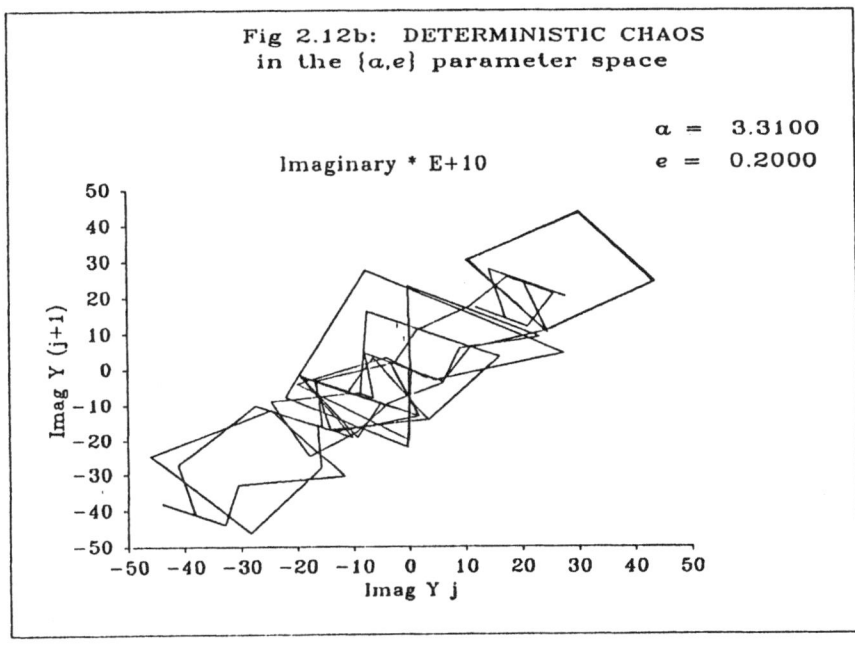

Fig 2.12b: DETERMINISTIC CHAOS
in the {a,e} parameter space

a =   3.3100
e =   0.2000

Imaginary * E+10

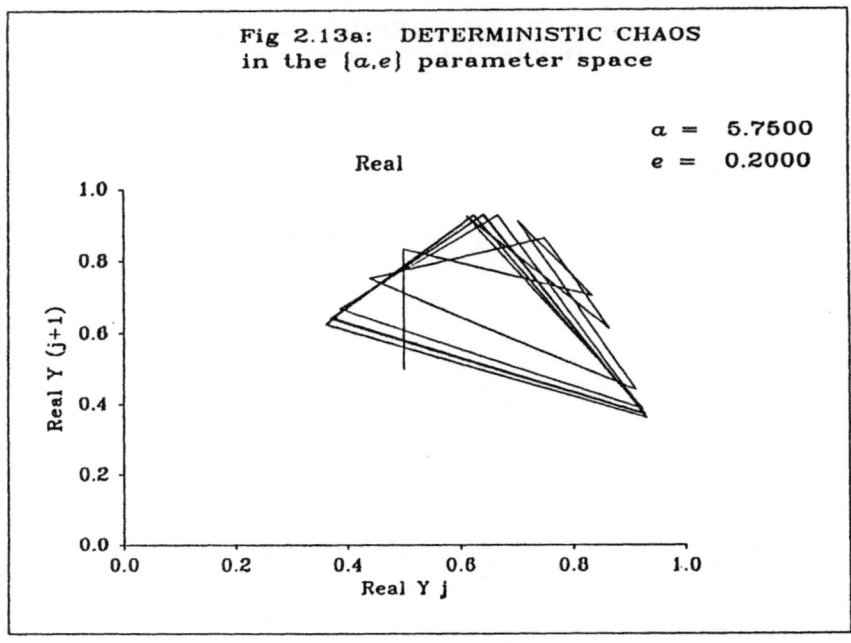

Fig 2.13a: DETERMINISTIC CHAOS
in the {a,e} parameter space

a =   5.7500
e =   0.2000

Real

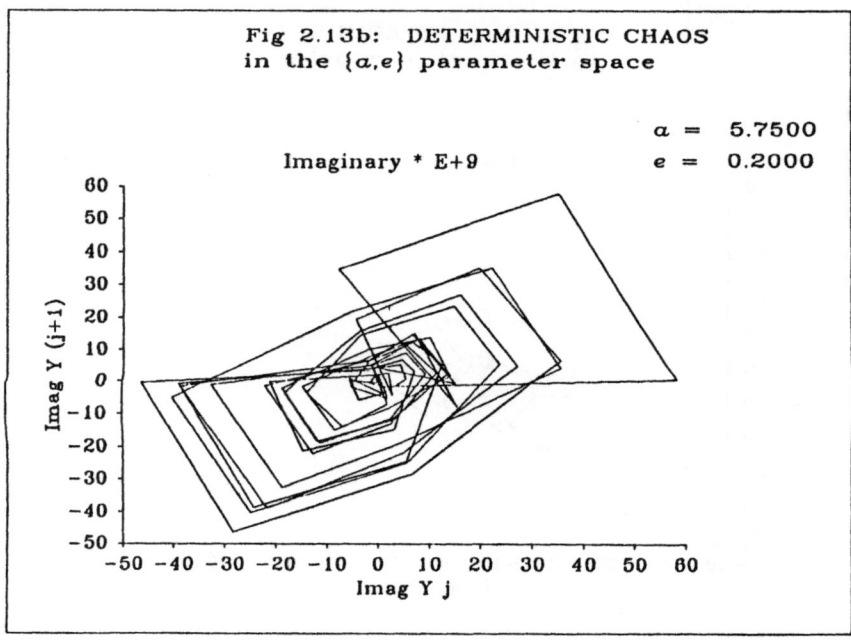

Fig 2.13b: DETERMINISTIC CHAOS
in the {a,e} parameter space

a =   5.7500
e =   0.2000

Imaginary * E+9

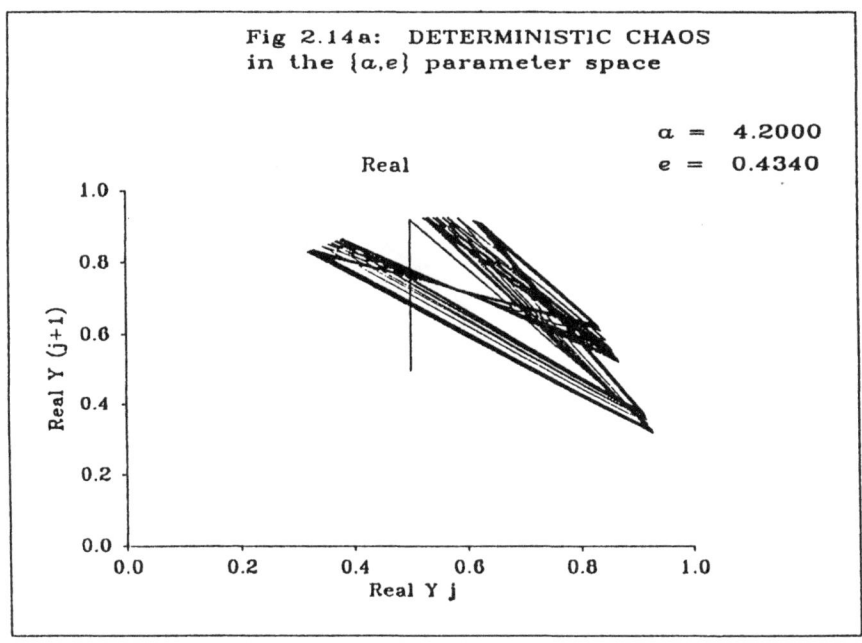

Fig 2.14a: DETERMINISTIC CHAOS
in the {a,e} parameter space

a =   4.2000
e =   0.4340

Real

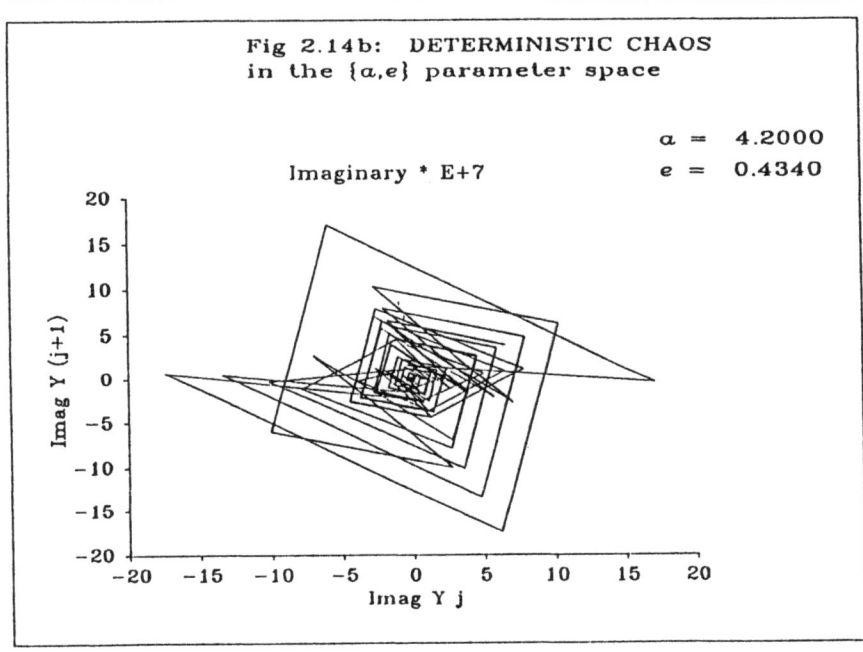

Fig 2.14b: DETERMINISTIC CHAOS
in the {a,e} parameter space

a =   4.2000
e =   0.4340

Imaginary * E+7

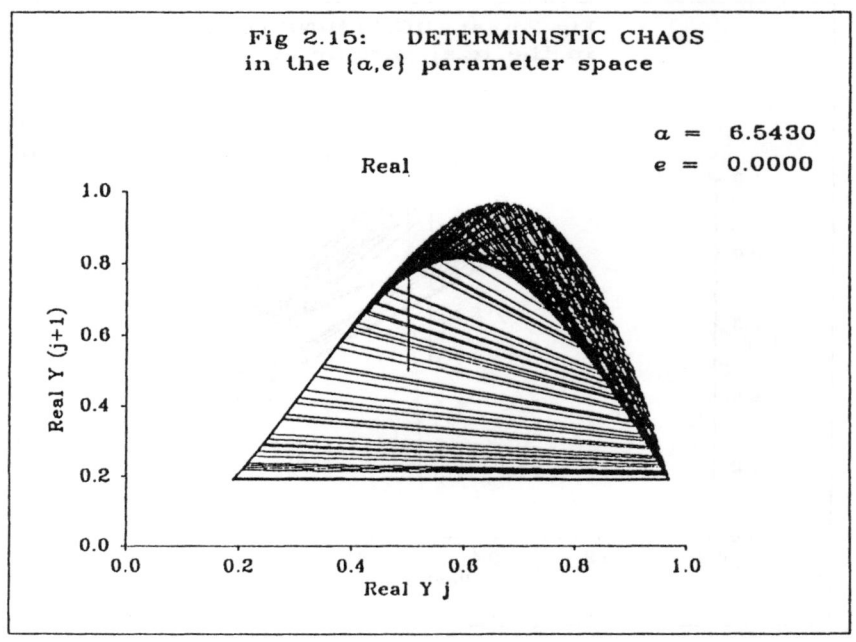

Fig 2.15:   DETERMINISTIC CHAOS
in the {a,e} parameter space

a =   6.5430
e =   0.0000

Real

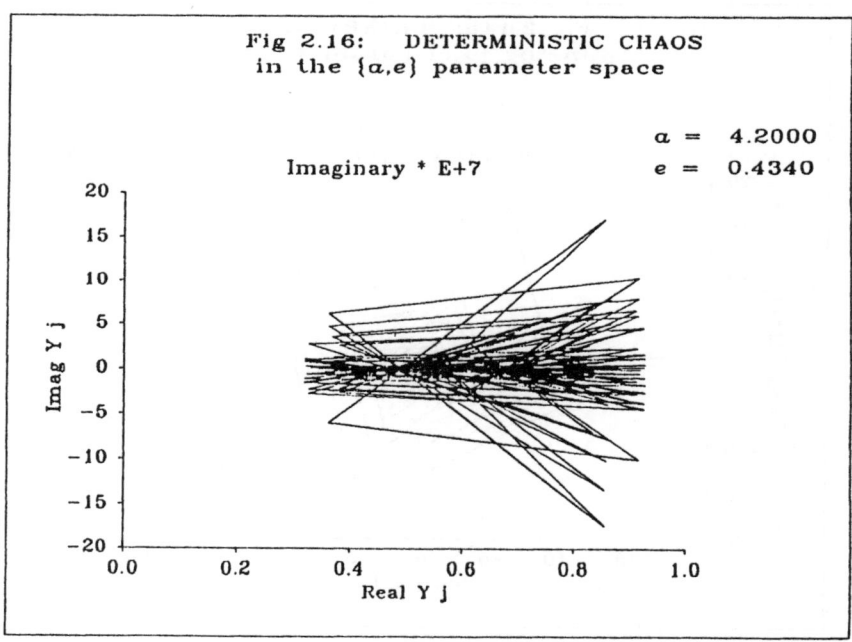

Fig 2.16:   DETERMINISTIC CHAOS
in the {a,e} parameter space

a =   4.2000
e =   0.4340

Imaginary * E+7

and $N(Y_i, r)$ is the number of points in a hypercube of side r, centred at the reference point Y_i (i is a coordinate set of one hypercube).

So $log[Q(r)]/log(r)$ is an estimate of D, called by some writers the "correlation dimension". This value is not necessarily an integer, and its value appears to depend upon the integer dimensionality of the embedding space; which must be greater. Some tentative estimates of the dimensionality of some attractors in this system suggest that they are between 1 and 3. Holzfuss and Mayer-Kress (1986) give methods that are in general computationally laborious and not of great empirical value such as to justify their precise determination.

For purposes of modelling human psychophysical data, the lower gain a values for which $\bar{\lambda}$ becomes positive may be more important[8], as they could be predictive of the degeneration of behaviour under acute overload. A model of the sort considered here is a potential bridge between normal sensory functioning and the unstable psychosensory phenomena associated with migraine and epileptic auras (Walsh, 1978, compare King and Boreham, 1983). Thus the first two or three regions of the Lyapunov graph for which $\bar{\lambda}$ becomes positive and the stable periodic windows that separate them may prove the most interesting, for studying clinical hypersensation.

A cautionary note must be added; there are at least three levels which can conceptually be distinguished when we consider a model with variable $\bar{\lambda}$. These are the analytic mathematics, the computer simulation, and the biological or psychological observable realization of the process. There is an intrinsic coarseness in the computer simulation such that very fine fluctuations in $\bar{\lambda}$ are either missed, or by local averaging reversed in sign, and the subsequent loss of information at the psychological level, which is inevitably even coarser, would be even greater, only the largest windows might be expected to have meaning in the psychopathology of behaviour.

Further work on the effects of noise on chaotic maps suggests some complications. The chaotic properties of a process imply a decay of serial correlations, or transinformation loss, in prediction. As Herzel and Ebeling (1985) comment, "low-dimensional chaotic systems are inevitably coupled to external fluctuations". The serial correlation of chaotic processes can be

---

[8] If it is known that the system has more than one dimension, then the Lyapunov dimension as defined by Frederickson *et al* (1983) is, for $j + 1$ dimensions,

$$D_L = j + \frac{\lambda_1 + \dots\dots + \lambda_j}{|\lambda_{j+1}|}.$$

A three-dimensional strange attractor must have Lyapunov exponents which are $+, 0, -$ and $\sum \lambda_j < 0$. For any bounded attractor which is not an equilibrium point, one $\lambda_j > 0$.

enhanced by external noise, and the noise amplifies some higher periodicities. If a system has two basins of stability within its parameter space, the boundary between these basins can become fractal instead of smooth as a result of long-delay terms in the generating recursion (e.g., putting $Y_{j-k}, k > 6$ in [2.2] instead of Y_{j-1}), which increases the probability of a system jumping between different dynamic modes (Aguirregabiria and Etxebarria, 1987). As Herzel and Pompe (1987) state "the local stability properties of a dynamic system are in general as complicated as the orbit itself ". This means that complicated residual errors, even if small, must arise in fitting the behaviour of any real psychophysical sensory channel to models as simple as [2.2]. though the presence of noise can in fact stabilize the system (Gwinn and Westervelt, 1986) and increase its predictability, and can also induce intermittency, which is a well-known feature of neurobiological networks.

References

Aguirregabiria, J. M. and Etxebarria, J. R. (1987) Fractal Basin Boundaries of a Delay-Differential Equation. *Physics Letters A, 122,* 241 - 244.

Arnold, L. and Wihstutz, V. (1986) Lyapunov Exponents: A survey. *In* Dold, A., and Eckmann, B. (Eds.) *Lyapunov Exponents. (Lecture Notes in Mathematics No. 1186).* Berlin: Springer-Verlag, pp. 1 - 26.

Argoul, F. and Arneodo, A. (1984) Lyapunov Exponents and Phase Transitions in Dynamical Systems. *In* Dold, A. and Eckmann, b. (Eds.) *Lyapunov Exponents.* (Lecture Notes in Mathematics No. 1186) Berlin: Springer-Verlag, pp. 338 - 360.

Cowan, J. D. (1968) Statistical Mechanics of Nervous Nets. *In* Cacaniello, E. R. (Ed.) *Neural Networks* . Berlin: Springer-Verlag.

Fiegenbaum, M. J. (1978) Quantitative Universality for a Class of Nonlinear Transformations. *Journal of Statistical Physics, 19,* 25 - 52.

Fiegenbaum, M. J. (1979) The Universal Metric Properties of Nonlinear transformations. *Journal of Statistical Physics, 21,* 667- 706.

Fischer, P., and Smith, R. (1985) (Eds.) *Chaos, Fractals and Dynamics (Lecture Notes in Pure and Applied mathematics No. 98),* New York: Marcel Dekker.

Fogelman-Soulié, F. (1986) Lyapunov Functions and their use in Automata Networks. *In* Bienenstock, E., Fogelman -Soulié, F., and Weisbuch, G. (Eds.) *Disordered Systems and Biological Organization* . Nato ASI Series Vol F20. Berlin: Springer-Verlag, pp. 85 - 100.

Fowler, A. C., Gibbon, J. D., and McGuinness, M. J. (1982) The Complex Lorenz Equations, *Physica, 4D,* 139 - 163.

Frederickson, P., Kapla, J. L., Yorke, E. D., and Yorke, J. A. (1983) The

Liapunov dimension of strange attractors. *Journal of Differential Equations, 49*, 185 - 207.

Freeman, W. J. (1979) Nonlinear Gain Mediating Cortical Stimulus - Response Relations. *Biological Cybernetics, 33*, 237 - 247.

Freeman, W. J. (1983) Dynamics of Image Formation by Nerve Cell Assemblies. *In* Basar, E., Flohr, H., Haken, H., and Mandell, A. J. *Synergetics of the Brain* . Berlin: Springer-Verlag.

Grassberger, P., and Procaccia, I. (1983) Characterization of Strange Attractors. *Physical Review Letters, 50*, 346 - 349.

Gregson, R. A. M. (1983) *Time Series in Psychology*. Hillsdale, New Jersey: L. Erlbaum Associates.

Gregson, R. A. M. (1984a) Invariance in Time Series Representations of 2-input 2- output Experiments. *British Journal of Mathematical and Statistical Psychology, 37*, 100 - 121.

Gregson, R. A. M. (1984b) Similarities between Odor Mixtures with known Components. *Perception and Psychophysics, 35*, 33 - 40.

Gregson, R. A. M. (1985) The Subjective Weber Function and Output Uncertainty in Nonlinear Psychophysics. Paper presented at the Osnabrück Mathematical Psychology Meeting, Nov. 1985. Universität Osnabrück, FRG.

Gregson, R. A. M. and Gates, A. (1985) Cross-modal Identification: Effects of Contingent Changes in the Stimulus Series. *Biological Cybernetics, 52*, 247 - 258.

Guckenheimer, J. (1980) One-dimensional Dynamics. *Annals of the New York Academy of Sciences, 357*, 343 - 347.

Gwinn, E. G. and Westervelt, R. M. (1986) Fractal basin boundaries and intermittency in the driven damped pendulum. *Physical review A, 33*, 4143 - 4155.

Hellman, R. P. and Zwislocki, J. (1964) Loudness function of a 1000-cps Tone in the Presence of Masking Noise. *Journal of the Acoustical Society of America , 36*, 1618 - 1627.

Herzel, H.-P. and Ebeling, W. (1985) The Decay of Correlations in Chaotic Maps. *Physics Letters, 111A*, 1 - 4.

Herzel, H.-P. and Pompe, B. (1987) Effects of Noise on a Nonuniform Chaotic Map. *Physics Letters A, 122*, 121 - 125.

Holzfuss, J., and Mayer-Kress, G. (1986) An Approach to Error-Estimation in the Application of Dimensional Algorithms. *In* Mayer-Kress. G. (Ed.) *Dimensions and Entropies in Chaotic Systems. (Springer Series in Synergetics, No. 32)*. Berlin: Springer-Verlag, pp. 114 - 122.

King, R., and Barchas, J. D. (1983) Theoretical Psychopathology: An Application of Dynamical Systems Theory to Human Behaviour. *In*

Başar, E., Flohr, H., Haken, H., and Mandell, A. J. (Eds.)*Synergetics of the Brain (Springer Series on Synergetics No. 23).* Berlin: Springer-Verlag, pp. 352 - 364.

Kornbrot, D. E. (1984) Mechanisms for categorization: Decision criteria and the form of the psychophysical function. *British Journal of Mathematical and Statistical Psychology, 37,* 184 - 198.

Kreuzer, E. (1987) *Numerische Untersuchung nichtlinearer dynamischer Systeme.* Berlin: Springer-Verlag.

Hao, B.-L. (1984) *Chaos.* Singapore: World Scientific Publishing Co.

Lorenz, E. N. (1963) Deterministic Nonperiodic Flow. *Journal of the Atmospheric Sciences, 20,* 130 - 141.

Mackey, M. C., and Glass, L. (1977) Oscillation and Chaos in Physiological Control Systems. *Science, 197,* 287 - 289.

Mandelbrot, B. B. (1980) Fractal Aspects of the Iteration of $z \mapsto \lambda z(1 - z)$ for Complex λ and z. *Annals of the New York Academy of Sciences, 357,* 249 - 259.

Marotto, F. R. (1982) The Dynamics of a Discrete Population Model with Threshold. *Mathematical Biophysics, 58 ,* 123 - 128.

May, R. M. (1976) Simple mathematical Models with very complicated Dynamics. *Nature, 261 ,* 459 - 467.

May, R. M. and Oster, G. F. (1976) Bifurcation and Dynamic Complexity in Simple Ecological Models. *The American Naturalist, 110 (974),* 573 - 599.

Mayer-Kress, G. (Ed.) (1986) *Dimensions and Entropies in Chaotic Systems. (Springer Series in Synergetics, No. 32.)* Berlin: Springer-Verlag.

Metropolis, N., Stein, M. L. and Stein, P. R. On Finite Limit Sets for Transformations on the Unit Interval. *Journal of Combinatorial Theory (A), 15,* 25 - 44.

Oseledec, V. I. (1968) A multiplicative ergodic theorem: Lyapunov characteristic numbers for dynamical systems. *Transactions of the Moscow Mathematical Society, 19,* 197 - 231. (in Russian).

Ott, E. (1981) Strange Attractors and Chaotic Motions of Dynamical Systems. *Reviews of Modern Physics, 53 (4), Part 1,* 655 - 671.

Price, I. R. and Gregson, R. A. M. (1988) Nonlinear Dynamics in a Complex Cubic One-Dimensional Model for Sensory Psychophysics. *Acta Applicandae Mathematicae, 11* 1 - 17.

Schuster, H. C. (1984) *Deterministic Chaos.* Weinheim: Physik-Verlag, 1984.

Schwartz, E. L. (1984) Spatial Mapping and Spatial Vision in Primate Striate and Infero-temporal Cortex. *In* Spillman, L. and Wooten, B. R. (Eds.) *Sensory Experience, Adaption, and Perception. Festschrift for Ivo Kohler.* New Jersey: L. Erlbaum Associates, pp. 73 - 104.

Shaw, R. (1981) Strange Attractors, Chaotic Behavior and Information Flow. *Zeitschrift für Naturforschung, 36a,* 80 - 112.

Singher, D. (1978) Stable Orbits and Bifurcation of Maps of the Interval. *SIAM Journal of Applied Mathematics, 35(2),* 260 - 267.

Verhulst, P. F. (1844) *Mémoires de l'Academie Royale de Bruzelles, 28,*1. *Cited in:* West, B. J. *An Essay on the Importance of Being Nonlinear.* Berlin: Springer-Verlag, 1985.

Walsh, K. W. (1978) *Neuropsychology: A Clinical Approach.* New York: Churchill Livingstone.

Werner, G. and Mountcastle, V. B. (1963) The variability of central neural activity in a sensory system, and its implications for the central reflection of sensory events. *Journal of Neurophysiology, 26,* 958 - 974.

Zeeman, E. C. (1976) Duffing's Equation in Brain Modelling, *The Institute of Mathematics and its Applications, 1976 (July),* 207 - 214.

Zeeman, E. C. (1977) *Catastrophe Theory: Selected Papers, 1972 - 1977.* Reading, Mass: Addison-Wesley.

3 A Recursive Loop System Using Gamma

The Γ recursion [2.2] is as it stands purely mathematical with interestingly complicated properties. However it is chosen because it may be amenable to interpretation as a model of mass action in a sensory system. Either it is a useful model, by subsuming the apparently diverse phenomena A1 to A19 within one process, or it is of a very limited value in pointing out differences between its behaviour and that of one or more of the human senses concerned with the experience of sensory intensity.

Figure 3.0 sets out the skeleton of how [2.2] is used as the equation of a recursive feedforward loop[1]. The input is a series $\{U_J\}$ of real positive scalars, $J = 1, 2, ..., N$. The observable output is the series $\{Y_{obs,J}\}$ whose generation in terms of the parameters (a, e, η) has already been defined. If Γ is a psychophysical system then the observables are the ordered pairs (U, Y). The creation and testing of a psychophysical model based on Γ is essentially reducible to a transfer function analysis of the $U \implies Y_{obs}$ mapping. The variables of the system are *both* the parameters (a, e, η) *and* the couplings W between $U, \Delta^1 U$ and a, e, η. There may be defined

---

[1] This idea of treating a sensory channel as a feedforward loop is not mere caprice, it is consistent with a skeletal representation of sensory-cognitive neurophysiology (Mishkin and Appenzeller, 1987) if the loop is taken as involving a recursive pathway through the sensory projection areas, then the amygdala or hippocampus (two alternative pathways), the diencephalon and the prefrontal cortex, and returning via the basal forebrain to the sensory areas.

constraints on $\Delta^1 Y_j$ which we may wish to postulate, and it is possible to inject gaussian random perturbations onto each of a, e, η. The number of alternative possibilities thus generated is very large, and there is no point in cataloguing or simulating them all. Some judgment concerning the psychological or psychoneurological plausibility of a mapping can be effected. Models which are in no sense identifiable in terms of location in a bounded subspace of $\{a, e, \eta, W\}$ which will support them are to be avoided. Some general comments on model identification must therefore be made.

Here for convenience the notation used is tabulated, to designate in sequence the observable and unobservable stages from input through to output:

$S = $ stimulus present (observable)

$U_J = $ quantified representation of the Jth stimulus property

((Re), +ive, observable)

$a = $ destabilizing parameter (Re)

$ie = $ oscillating parameter (Im)

$Y_j = $ internal loop parameter (Re, Im)

$j = $ recursion , $1, ..., \eta$

$\eta = $ dwell time in loop

$W = $ couplings of $U_J, \Delta^1 U_J$ onto a, e

$Y_{obs.J} = Y_{J.\eta}$ quantified aspect of output

$\mathcal{R} = $ response (may be degraded Y_{obs}) (observable)

Inputs to Γ and Identification

The input series $\{U_J\}$ has to have some defined sequential properties, and the two obvious cases are i.i.d. (*identically independently distributed*) series, and strictly periodic series. One could for example construct systems in which

$$\Delta^1 U_J = f(\sum_{k=0}^{m} w_k \cdot Y_{obs.J-k}) \qquad [3.1]$$

and solve for $\{w_k\}$ as a model of contingent adaptive behaviour, but initially it is clearer only to consider systems in which the outer loop of U, Y_{obs} is open. $\{U_J\}$ is specifiable in this context by its autoregressive spectrum, and associated entropy. In all the variants of System Γ to be studied here the mapping $U \Longrightarrow a$ (and note, *not* $U \Longrightarrow Y_{obs}$) is assumed. This is

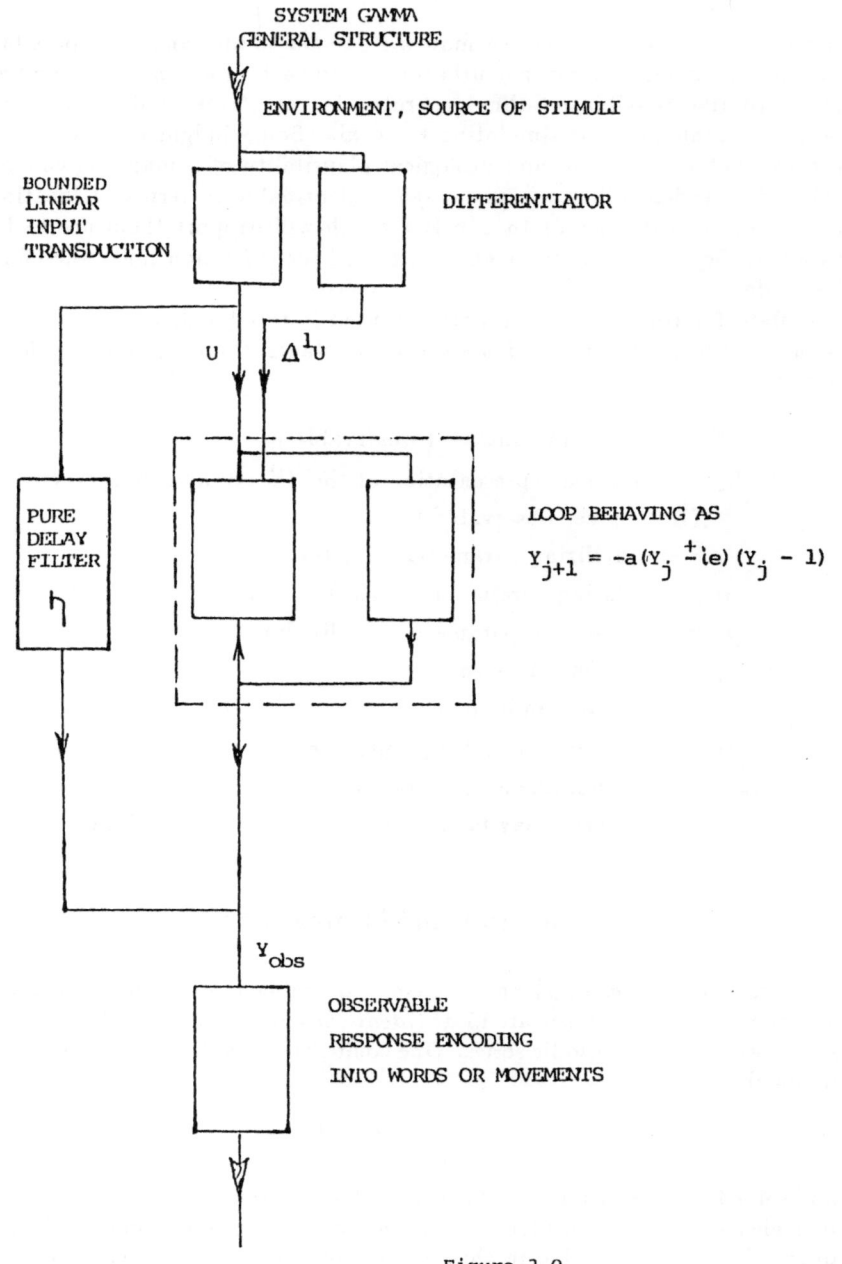

Figure 3.0

axiomatic, and provides for studying system behaviour where increasing input induces greater system instability and noise, but in a manner which involves nonlinearity.

For simplicity consider the case where $U \Longrightarrow a$, with e and η fixed. This elsewhere (Gregson, 1985) we have coded as Γ V 7. The system can have additional noise injected at various points; as second-order noise on the mapping $U \Longrightarrow a$, on the parameters e, η and as limits on $\Delta^1 \epsilon$ or $\Delta^1 Y$ for $J, J + 1$.

The identification of a psychophysical system is in linear modelling reported in the form of estimating the maximum likelihood values for the system parameters, given the model structure. One assumes, for example, that

$$Y = \alpha(U - U_0)^\beta + \epsilon \qquad [3.2]$$

and determines β from an experimental regression of $logY$ on $logU$, if necessary estimating U_0 to make the regression approximate to a straight line[2]. The implicit assumptions are

$$\epsilon \sim \mathbf{N}(0, \sigma_\epsilon) \qquad [3.3]$$

$$cov(U, \epsilon) = 0 \qquad [3.4]$$

$$cov(\epsilon_j, \epsilon_{j-k}) = 0 \qquad [3.5]$$

and that no terms in UY exist. Nothing is assumed about $cov(U_j, U_{j-k})$ because trials are taken to be sequentially independent. Tabulation of β values for a diversity of sensory continua is the sort of psychophysical analysis associated with [3.2]. As [3.3] to [3.5] are both false and unnecessary, and as we can legitimately have considerable interest in $cov(U_j, U_{j-k})$ is real-world situations (the biological survival of a higher organism depends on its relative ability, as compared with other species, to extrapolate a little into the future), given a stimulus series, [3.2] is of but slight value; it is an approximation to a gain function and not a process model. Why [3.2] sometimes works as a useful first approximation is reviewed by Gregson (1983). It follows that if we use Γ to create gain functions then they should

---

[2] It is one of the persistent myths of psychophysics that this equation does yield a good straight-line fit in log-log plots, and that it is due to S. S. Stevens. The equation actually originates with a Belgian, J. A. F. Plateau, in 1872, and usually fails to fit data, the departures being consistently like a weak CNO form. An interesting extension of this finding is reported by Müller (1987) where the departure is shown, in auditory psychophysics, to hold within subranges of a two-stage (coarse then fine) category scaling procedure.

algebraically not be too different from [3.2] over a limited range of U values. This turns out to be very easy to achieve, without invoking the stochastic term ϵ or its analogue as noise injected into Γ.

However, the slope of a fitted psychophysical gain function in [3.2] is dependent on α, β, and σ_ϵ, and posterior confidence intervals for $\hat{\alpha}, \hat{\beta}, \hat{\sigma}_\epsilon$ can be obtained given a prior probability of the appropriateness of [3.2], also provided that stationarity assumptions hold and that information on the dynamic transfer function $U_J \mapsto Y_J$ is not to be fitted. In short, the dynamics of $U \implies Y$ are filtered out and lost irretrievably. There is not an exact parallel to *parameter* estimation in nonlinear models, because nonlinear models and some biological systems are robust in their parameters; they exhibit what is called Andronov stability. This means one can make some changes in their internal parameter values and leave the gain function almost unperturbed, in other subregions of the parameter space tiny changes in parameters induce large and sometimes disastrous changes in output.

Whilst nonlinear systems are not uniformly sensitive to changes in their internal parameters, at the same time in order to identify what internal structure such a system has there have to be some properties of the input-output mapping which are uniquely associated with subregions of the system parameter space, in our case with $\{a, e, \eta, W\}$. If such properties do not exist then one has no insight, dynamically, into where the system is at present and to where it is going next.

There are at least five properties of the output, given that we know the input in the sense of [3.3] to [3.5] and the history U_0 to U_J, whose existence can be used collectively to assist in identification. These properties can be used to impose limits on the plausible values of internal parameters. To use a bizarre metaphor, some systems are semi-opaque black boxes. Parameter identification is necessary for the prediction of behaviour, and parameter identification in turn presupposes some current system structure identification.

The five properties, all of which can be quantified, are

1: The input-output (U, Y) topology.

2: The consequences in Y of small changes in U and its rate of change $\Delta^1 U$.

3: The degradation in the entropy $D_H = H(Y) - H(U)$.

4: The existence of local outbursts and gaps in the response series, and the distribution of interoutburst intervals in time.

5: The autoregressive spectrum of the output, $\{v_k \mid Y\}$ which itself may or may not be amenable to a linear time series representation.

Each of the properties 1: to 5: already has an extensive technical literature in systems theory.

System Γ is not advanced as a Short Term Memory (*STM*) model, but because it is recursive implies that the system has a memory of duration η for one point-input. Putting $Y_{obs.\eta}$ into LTM and retrieving it is not covered by this model, so only $S - \mathcal{R}$ behaviour with reasonably short response latencies is predictable. Further, the observed latency of a response is not the same thing as η; only if the time to process $S \Longrightarrow U \Longrightarrow a$, and $Y_{obs} \Longrightarrow \mathcal{R}$ are almost fixed and independent of $U, \mathcal{R}$ can η be used as a metric predictor of reaction times .[3] There is nonetheless an analogy with STM models based on neural network properties (Ellias and Grossberg, 1975), provided that we carefully distinguish between processes which support contrast or discrimination between pattern stimuli but only have unstable limit cycles, and processes as here which represent a single stimulus (which may be a vector) and can have stable limit cycles or go transiently into chaos.

There is a possible counterinstance to this argument; we can cascade Γ loops, using $_1Y_{obs.\eta}$ as input to $_2a$, $_2Y_{obs.\eta}$ as input to $_3a$, etc. The final relationship for $n - 1$ cascades $k = 1, \ldots, n$ is $U \Longrightarrow_n Y_{obs.\eta}$. ($\eta$ is not necessarily equal for all $_kY_j$) which will in some cases have a gain function that does not at all resemble any single Γ loop process, and has an output time series resembling that of a bilinear process, with transient explosions (Gregson, 1983, sect. 2.5, gives examples). If $_n\Gamma$ is a valid model it would be expected to be able both to reflect the S,R relations and to have relatively very big response latencies, assuming that the comparison is n cascades for fixed η, and not $n\eta = const.$, which holds if the process has a choice between dwelling a long time in one loop or a very short time in each of a series of loops in cascade.

There is some interrelation between limit cycle activity in a network, and the form of input function that will facilitate pattern selection and storage. A consideration of alternatives by Ellias and Grossberg (1975) indicated that in some cases STM storage is facilitated by a sigmoid input function, because for pattern selection and representation an input which is faster-than-linear levelling off later is optimal. That argument rests on some reasonable assumptions about connectivity in neural networks with excitatory foci and inhibitory surrounds.

---

[3] The notion that the process time outside the Γ loop should be independent of input and output values relates to conditions for the identifiability of series-parallel and parallel stochastic systems reviewed by Townsend and Ashby (1983). Treating η as approximately constant, but not a reaction time, has an interesting parallel in the empirical results of Pöppel and coworkers, who demonstrate some generality for an internal discrete time sampling mechanism, with a resolution not finer than about 30 msecs (Pöppel and Logothetis, 1986, Madler and Pöppel, 1987).

Taking both the sigmoid input-output and the limit cycle property as desired system specifications, it is required to show first by simulation in a simple case of Γ that both can be generated under reasonable but not necessarily global parameter values.

Whilst psychophysical data often generate a CNO-like psychometric function, they also produce other forms which slightly resemble truncated segments of a CNO. For example a range of results in visual perception, which resembles comparable cases in hearing, is given by Rodenburg, Maas and Stassen (1981). It is therefore not only necessary to reproduce a sigmoid gain function, it is necessary to show that with small parameter changes the shape of the function will turn into one or other of forms that have been empirically observed. In doing this one may also produce forms that have never yet been seen; their status is ambiguous until real data confirm or refute their plausibility.

The Γ V 7 variant is defined as
$$0 < U < 1, \qquad , a_{min} < a < 4$$
$$e \text{ fixed}, \ \eta \text{ fixed}, \ Y_0 = .5, \epsilon \qquad (\text{complex})$$
$$a = a_{min} + (3.99 - a_{min}) \cdot U$$

The constraint $a < 4$ is not mathematically necessary (whereas a constraint on ae is mathematically necessary to avoid explosions) but is biologically plausible and simplifies matters a little. For simulation in Figures 3.1 to 3.8, which plot the gain function $Y_{obs}.{}_{10}(\text{Re})$ against U (both normalised separately in the range 0-50 for graphing purposes only) for 50 U values with U distributed i.i.d. rectangularly, $\eta = 10$. It is possible to constrain the mapping of $U \Longrightarrow a$ within any subrange inside $2 < a_{min} < a_{max} < 4$ and produce segments of the gain function topology in that manner. This leads to the possibility of exploring stimulus range effects in transposition experiments. Figure 3.5 shows an ogive in which the curvature is different below and above the point of inflexion, which thus resembles some empirical results cited in Chapter 1.

The nonlinear properties of sensory neurophysiological systems, and in particular olfaction (Freeman, 1975, 1979, 1983, Marczynski, 1983) suggest very strongly that the *normal* mode of signal processing is one in which an input burst, extended briefly in time, is converted into a limit cycle pattern in which stimulus properties are phase-encoded, but in a space-dependent fashion and not a time-dependent one. There is also a biological need for the system to have a continuous representation of internal noise as a background component, without which it collapses onto a point, as in a state of anaesthesia (Freeman, 1987). We are thus led to ask here, can the theoretical system Γ produce limit cycles both internally and in its observable output from an input series U which is almost constant but with random uncorrelated (i.i.d.) second-order perturbations of its parameters

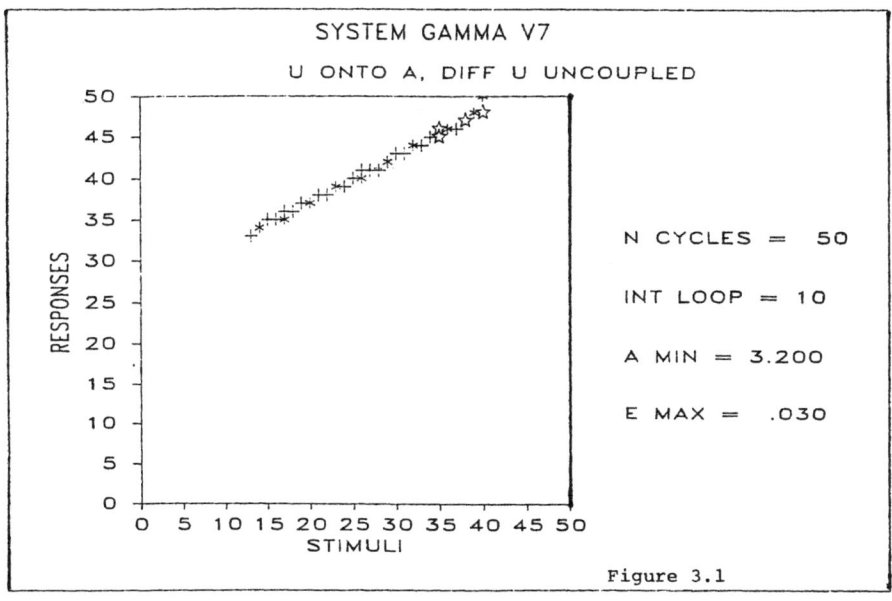

Figure 3.1

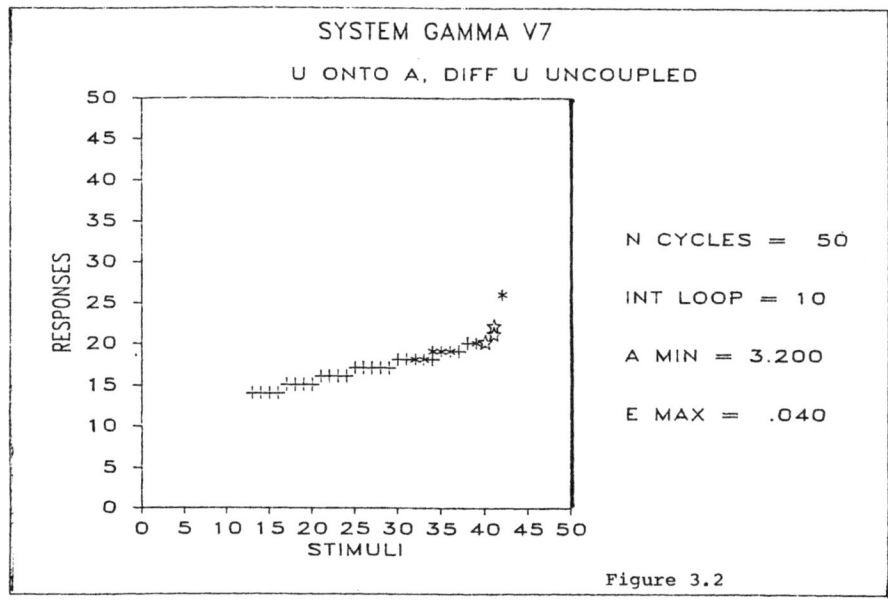

Figure 3.2

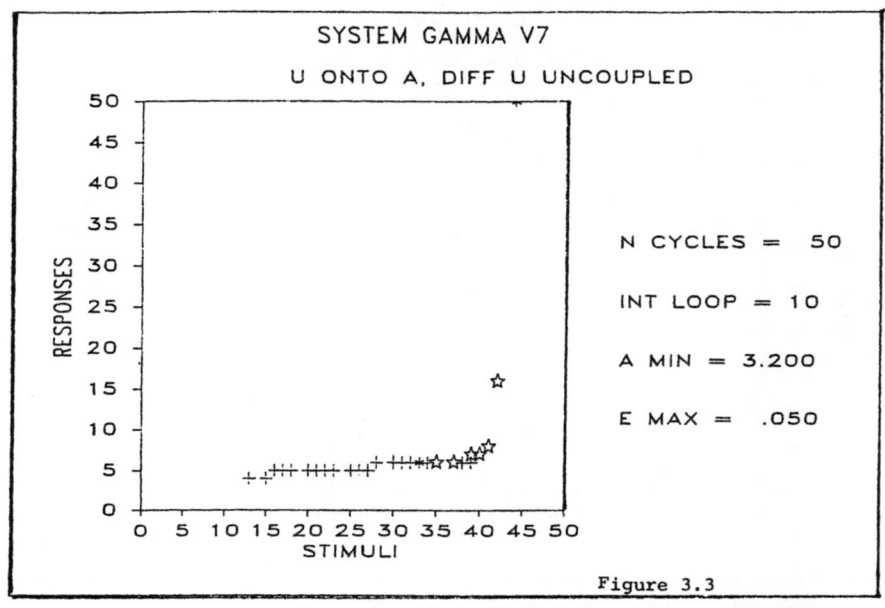

Figure 3.3

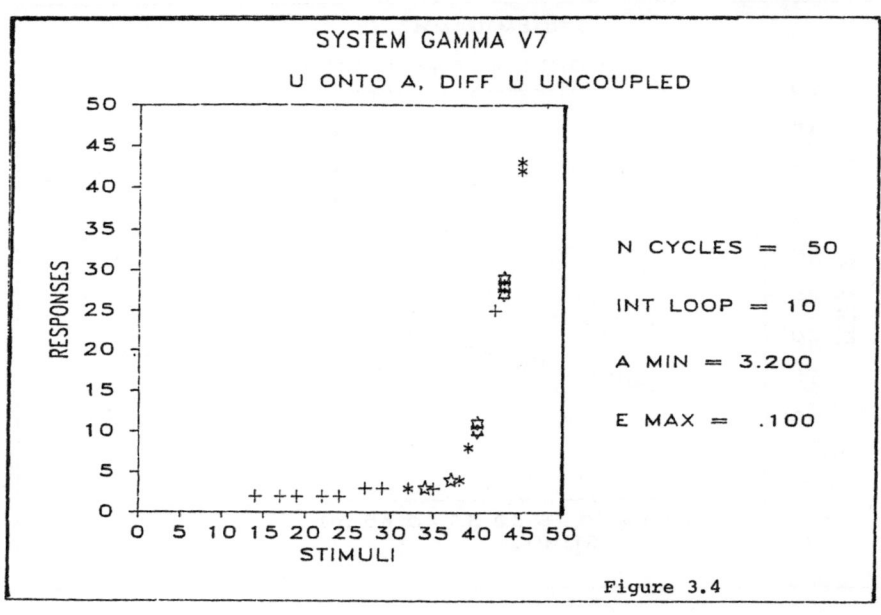

Figure 3.4

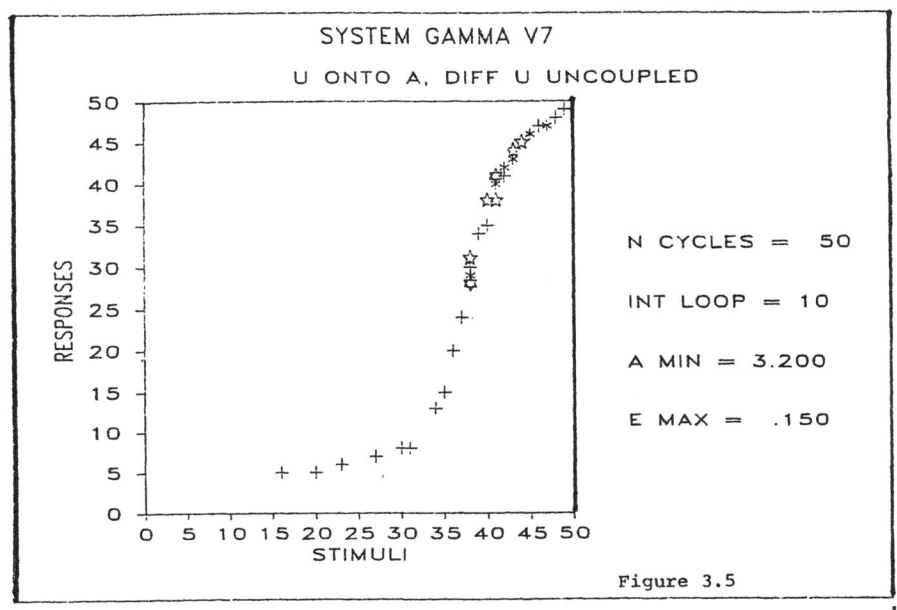

Figure 3.5

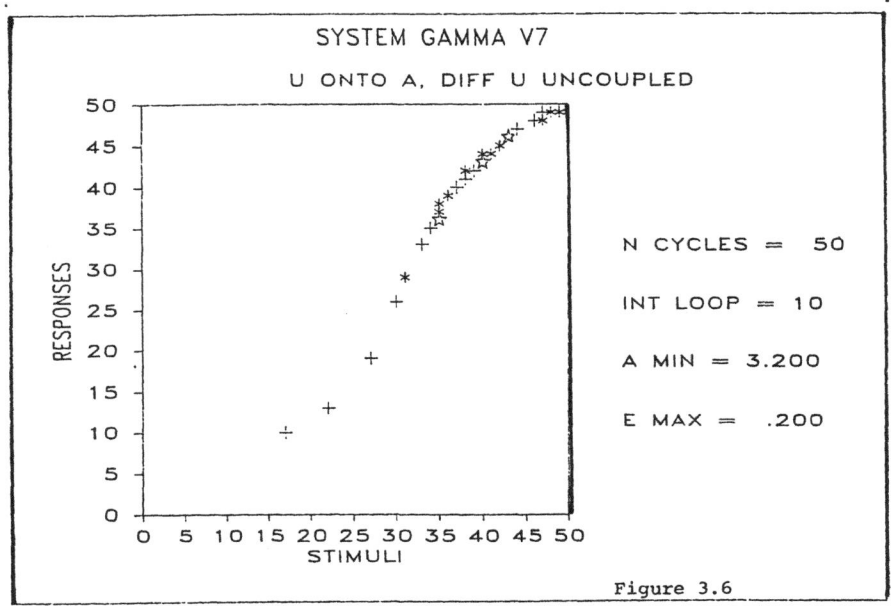

Figure 3.6

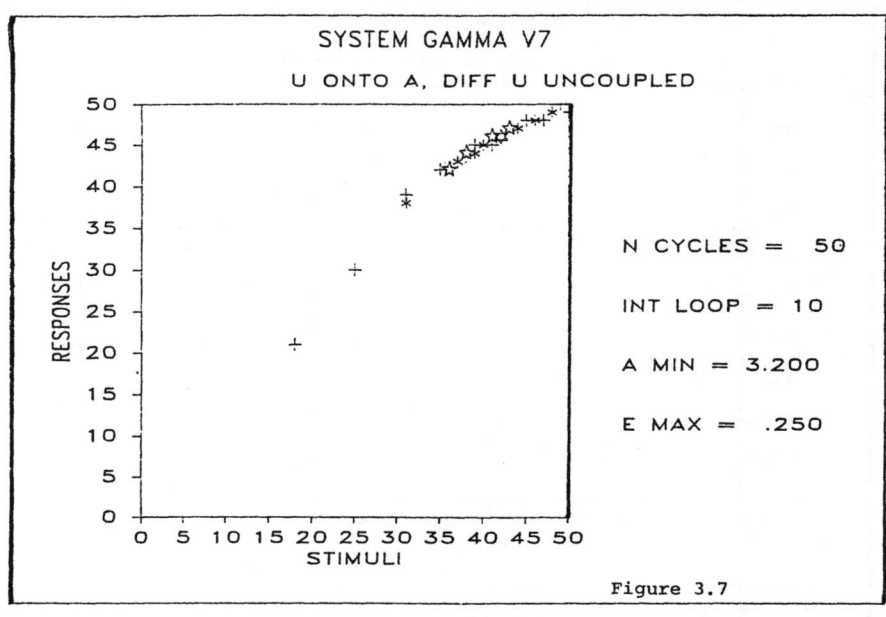

Figure 3.7

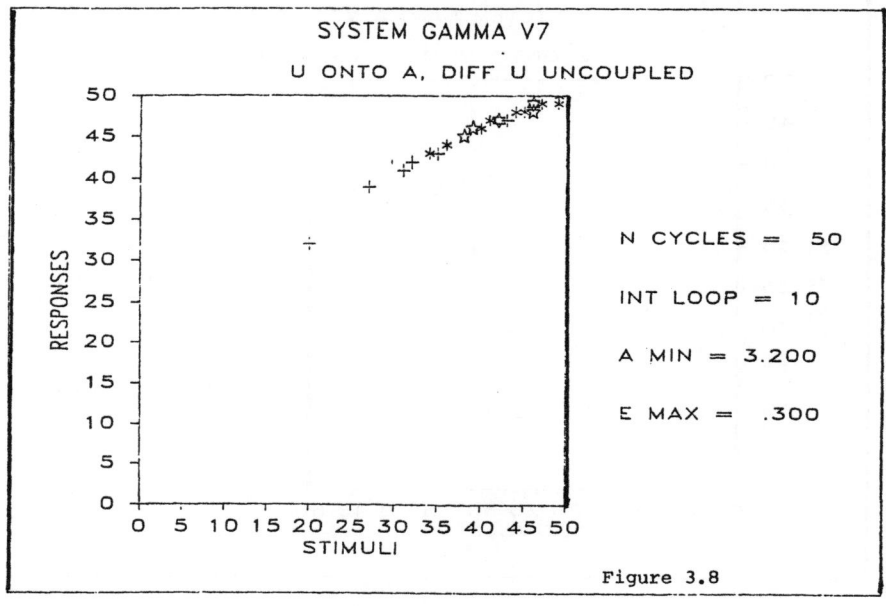

Figure 3.8

a and *η* ?

If a sensory system is internally in a limit cycle mode, it is required to find out if there is a predictable relationship between the quasiperiodicity of the internal physiological limit cycle and the externally observed response sequences, which themselves may also cycle at a different frequency. Unless it is possible to generate limit cycles without periodic input, and within a parameter range compatible with that viable for simulation of other psychophysical phenomena, then the model is not sufficient.

Inducing a range of Y_{obs}(Re) from a range of U, on either systematic or random input sequences, can often produce in Γ with fixed e and fixed $η$, and a range of a, an input output function like Figures 3.1 to 3.8 with a sigmoid and monotone gain function showing no signs of hysteresis. That is, the $S - R$ mapping can be almost $1 \rightarrow 1$, though cycling requires that it should be $1 \rightarrow many$, but the extent to which a response takes repeatable multiple values would depend upon the phase-locking of the inner cycling and the output observations in real time. If the latter is of a very low frequency compared with the inner cycling, and not phase-locked, we might expect a smudging of responses to a given stimulus which could fortuitously resemble a gaussian error distribution.

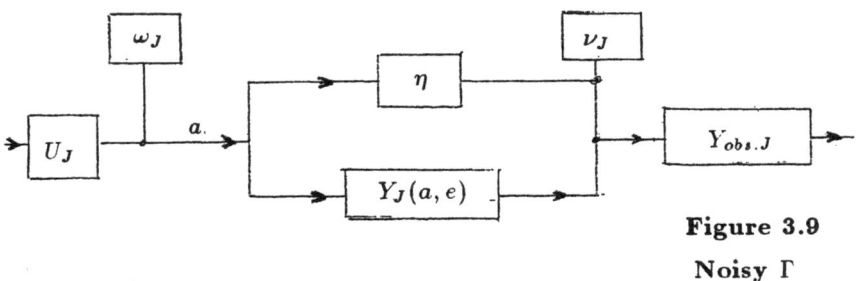

Figure 3.9

Noisy Γ

Consider Figure 3.9; noise is injected into V 7 upstream of a and into the pure delay path for $η$. The uncorrelated random noise sources are $ω_J$ and $ν_J$ as

$$ω_J = \mathbf{N}(0, σ_ω) \qquad [3.6]$$

$$ν_J = \mathbf{N}(\bar{ν}, σ_ν) \qquad [3.7]$$

In simulations the values $U = 3.2$, $σ_ω = .02$, $η = 5$, $\bar{ν} = 2.5$, $σ_ν = 1$ have been used. All this represents a noisy input source process with variable delay, but with fixed e.

Simulating a run of 100 U values with a rectangular probability distribution, the transfer function of $U \mapsto Y_{obs}$ (Re) can be fitted by an ARMA model **M1**, with the usual p, d, q notation of orders[4] by $p(3, 5, 6, 7, 8, 9)$, $d(0), q(1, 2, 3)$ with the cross-correlation between residuals $= .03$. The associated linear transfer function coefficients v_k are graphed in Figure 3.10. The striking feature of these v_k is their quasi-periodicity. In root locus form this implies a conjugate pair nearly on the imaginary axis; the irregularities and slight damping suggest at least the presence of a second conjugate pair of roots and a single negative real root. It is known that limit cycles can be associated with a conjugate pair moving over slightly into the righthand (positive, unstable) halfplane.

A model **M2** of inferior fit (the cross-correlation between residuals is .25) but simpler structure is $p(3, 5, 7, 9), d(0), q(1, 2, 3)$ which has a rapidly decaying and then oscillating transfer function. This suggests that a random input series is on the edge, dynamically, of generating limit cycles in the observable output. As Haken (1983) has observed, the brain as a system would have to operate at the limits of stability, so this result is not implausible in a model.

Table 3.1

Roots of the Transfer Functions for Models **M1** and **M2**

M1(Re)	M1(Im)	M2(Re)	M2(Im)
.704	-	.093	± .408
-.067	± 1.057	-.177	± 2.140
.004	± .169	.041	-
-1.113	± 5.631	-12.300	-
.078	± 2.083	-.074	± .918
-.378	-	-.729	-

From Table 3.1 it appears that the roots of the better fitting model **M1** include a dominant conjugate pair in the lefthand halfplane, indicating slow convergence, and two secondary conjugate pairs on or to the right of the Im axis. For the inferior model **M2** which converges more rapidly there are two negative real roots and two conjugate pairs near to the Im axis; the maximum delay component (Robinson, 1983) within the unit circle being in the righthand halfplane. This is graphed in Figure 3.11.

---

[4] p is the order of the autoregressive components of a time series model, d is the order of differencing, and q is the order of the moving average components.

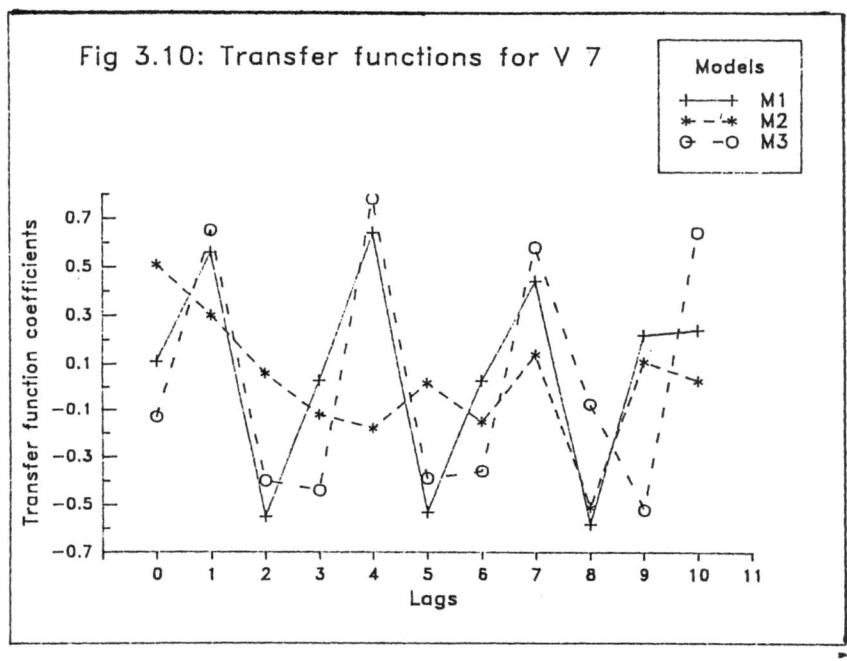

Fig 3.10: Transfer functions for V 7

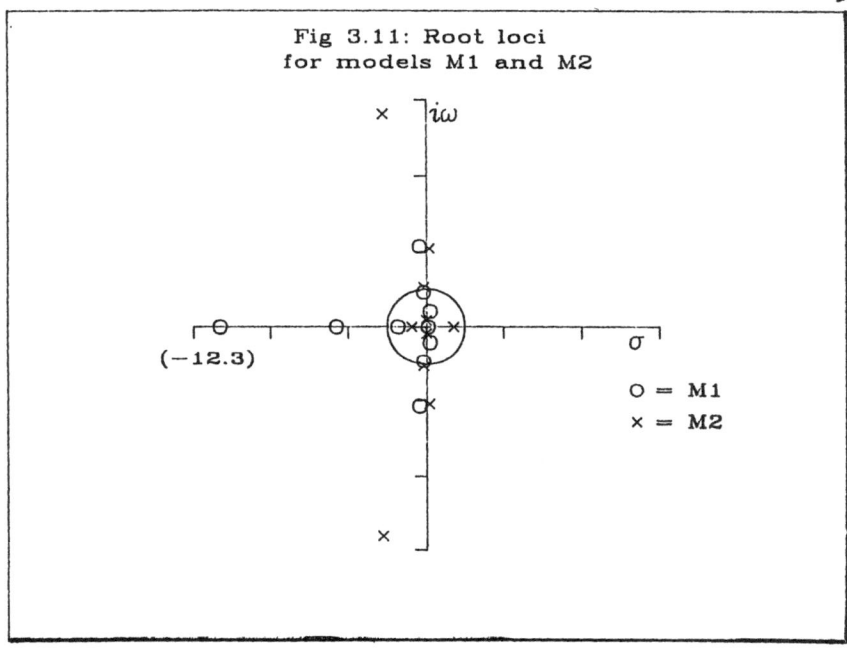

Fig 3.11: Root loci
for models M1 and M2

Whilst it would seem that the obvious and easiest way to induce limit cycling from random perturbations in the internal parameters is to induce noise in the loop delay parameter η, this may not be the only way. It is reasonable to consider first the effects of varying η as we would not expect that a biological system could or would need to maintain its internal timing exactly constant. Such timing is affected by the neurohormonal environment of the network and this can vary. Within limits the System Γ is parameter robust in η with respect to the input-output relations if $\eta > 5$, which is what has been assumed for these simulations.

It is interesting to note that the problem of deciding when a recursive sensory system should stop is a venerable one, and Herbart (1812) in one of the first examples known of the use of differential equations in sensory psychology (anticipating Fechner's psychophysics by nearly half a century) commented that

> "Aber num liegt allerdings die Frage in der Nähe: warum denn nicht die blosse Dauer den Grad der Wahrnehmung erhöhe ? Warum nicht längeres Hören den Ton verstärke, längeres Sehen die Farbe erhelle ?"

If η is held fixed and second-order random variation in e is allowed, again the system can show a tendency to limit cycle dynamics, but with different periodicity (see Figure 3.10; **M3**) and a dominant unstable oscillatory component. It is otiose to push the analysis further, precisely because to do so would involve a consideration of the covariance of random perturbations on different parameters within the system. There exists no information to guide conjectures on this point.

References

Ellias, S. A., and Grossberg, S. (1975) Pattern Formation, Contrast Control, and Oscillations in the Short Term Memory of Shunting On-Center Off-Surround Networks. *Biological Cybernetics, 20,* 69 - 98.

Freeman, W. J. (1975) *Mass Action in the Nervous System.* New York: Academic Press.

Freeman, W. J. (1979) EEG Analysis gives Model of Neuronal Template-Matching Mechanisms for Sensory Search with Olfactory Bulb. *Biological Cybernetics, 35,* 221 - 234.

Freeman, W. J. (1983) Dynamics of Image Formation by Nerve Cell Assemblies. *In* Başar, E., Flohr, H., Haken, H., and Mandell, A. J. *Synergetics of the Brain.* Berlin: Springer-Verlag.

Freeman, W. J. (1987) Simulation of chaotic EEG patterns with a dynamic model of the olfactory system. *Biological Cybernetics, 56,* 139 - 150.

Gregson, R. A. M. (1983) *Time Series in Psychology* . New Jersey: L. Erlbaum Associates.

Gregson, R. A. M. (1985) The Subjective Weber Function and Output Uncertainty in Nonlinear Psychophysics. Paper presented at an *Osnabrück Mathematical Psychology Meeting* , Universität Osnabrück, FRG, November, 1985.

Haken, H. (1983) Synopsis and Introduction. *In* Başar, E., Flohr, H., Haken, H., and Mandell, A. J. *Synergetics of the Brain* . Berlin: Springer-Verlag.

Herbart, J. F. (1812) Psychologische Untersuchung ueber die Staerke einer gegebenen Vorstellung, als Function ihrer Dauer betrachtet. *Reprint in* Hartenstein, G. (Ed.) (1889) *Johann Friederich Herbart's Schriften zur Psychologie, Dritter Theil. Kleinere Abhandlungen zur Psychologie.* Hamburg: Verlag von Leopold Voss.

Madler, Ch. and Pöppel, E. (1987) Auditory Evoked Potentials Indicate the Loss of Neuronal Oscillations during General Anaesthesia. *Naturwissenschaften, 74,* 42 - 43.

Marczynski, G. T. (1983) Algorithm for calculating theoretical probabilities of patterns of sequential inequality testing.*International Journal of Bio-Medical Computing, 14* , 463 - 486.

Mishkin, M., and Appenzeller, T. (1987) The Anatomy of Memory. *Scientific American, 256,* 62 - 71.

Müller, F. (1987) *Skalierung und Bezugssystem der Tonheit.* Inaugural-Dissertation; Würzburg: Julius-Maximilians-Universität zu Würzburg, Federal Republic of Germany.

Plateau, J. A. F. (1872) Sur la mésure des sensations physiques, et sur la loi qui lie l'intensité de ces sensations á l'intensité de la cause excitante. *Bulletin de l'Académie Royale Belgique, 33,* 376 - 388.

Pöppel, E. and Logothetis, N. (1986) Neuronal Oscillation in the Human Brain. *Naturwissenschaften, 73,* 267 - 268.

Robinson, E. A. (1983) *Multichannel Time Series Analysis (2nd Edn.)* Houston, Texas: Goose Pond Press.

Rodenburg, M., Maas, A. J. J., and Stassen, H. P. W. (1981) Thresholds for the Perception of Rotation: Variability, Psychometric Curves, and Comparison with Hearing Thresholds. *Biological Cybernetics, 42,* 23 - 28.

Townsend, J. T. and Ashby, F. H. (1983) *The Stochastic Modeling of Elementary Psychological Processes.* Cambridge: Cambridge University Press.

4 A Subjective Weber Function and Output Uncertainty

The original form of what is known commonly as *Weber's law* states that $\delta S/S = k$, where S is measured in stimulus (physical) units, and δS is an increment on S with a defined associated probability value, usually .5 by convention, of difference detection between S and $S + \delta S$. It is thus a venerable measure of response uncertainty as a function of stimulus intensity, and implies that the response uncertainty increases approximately linearly with the stimulus input, where that uncertainty is measured in stimulus units. It is not actually a psychophysical equation, as only physical measures appear in it, and both sides of the equation are dimensionless.

However, on the response side Eisler, Holm and Montgomery (1979) showed that we can define a subjective Weber fraction $\delta \mathcal{R}/\mathcal{R} = k'$, implying that the uncertainty of the response, in response units, increases with the mean response in a manner which is also monotonic, analogously to the original Weber formulation. This subjective Weber function is written in sensation units, and again is dimensionless. It is worth noting that in fact if either the Weber function or the subjective Weber function is written with second-order corrections for the fact that k is actually a function of S, and k' a function of $\mathcal{R}$, (empirically k' appears in some circumstances to vary parabolically with $\mathcal{R}$), then the expressions may cease to be dimensionless. In short, odd things can be observed to happen at the extremes of S for k, and at the extremes of $\mathcal{R}$ for k'. The empirical distinction between k and k' is a central concern here, together with the close relation between $\delta \mathcal{R}$ and associated measures of the imprecision of $\mathcal{R}$; $var(\mathcal{R})$ and $\mathbf{H}(\mathcal{R})$, the information or uncertainty in responses. The behaviour of Y_{obs} is examined in terms of its associated uncertainty.

The output Y_η of the nonlinear recursive process in Γ has been defined (Chapter 2) as a complex variable $Y_{obs}(\text{Re},\text{Im})$, and as has been noted the behaviour of $Y(\text{Re})$ and $Y(\text{Im})$ can be very different in their dynamics over some regions of the parameter space $\{a, e, \eta\}$. The derivation and form of Figures 4.1 and 4.2 amplify this comment.

In these plots, where $+$ indicates a mean output and $*$ the corresponding variance, the generation of the local variance for each value of a was created by varying e and η over a neighbourhood lattice spanning $7 \times 7 = 49$ e, η points in all. Averaging $Y_{obs}(\text{Re},\text{Im})$ over this local region of the parameter space yields local means and variances separately for both $Y_{obs}(\text{Re})$ and $Y_{obs}(\text{Im})$ for a fixed a. The behaviour of these four functions of a requires separate consideration. $Mean[Y_{obs}(\text{Re})]$ is a positively accelerated function of a (and hence of U, given that a is simply a linear transformation of U as assumed by W) over the range $1.5 < a < 3.95$. Below $a = 2$ there is, of course, virtually no response, and values above $a = 4$ have not been explored in this section. The positively accelerated form in Figure 4.1 matches commonly found $S - \mathcal{R}$ mappings, including for example, ones found for category scales by Montgomery and Eisler (1974). The associated $var[Y_{obs}(\text{Re})]$ rises and falls, in a parabolic fashion, the actual rises being an order of magnitude less. It is of interest here that the ascending limb of the parabola is almost straight for a range between the extremes; in other words k' is approximately constant. However, the skew parabolic turning-over is also found in scaling; again Montgomery and Eisler (1974) illustrate cases of an almost-symmetric parabola over a stimulus range that could reasonably be mapped into the values 20 - 50 in the arbitrary units of the graphs shown here. Ekman and Künnapas (1969) and Gregson (1976) reported what Gregson there termed a *banana-shaped* distribution of response variances against response means over the closed interval of 0,1 of similarity judgments; this term *banana-shaped* describes a skewed parabola. The Γ model is the first instance we know of a theoretical prediction of this skew parabolic form. So far it has not been derived readily from stochastic perturbation of similarity models (Gregson, 1975) that fit well the psychophysical relations between stimulus properties and mean similarity responses; the important point here is that the prediction follows naturally from Γ with no additional assumptions.

In Figure 4.2 the $Y_{obs}(\text{Im})$ values are $10^{-8} \times \text{Re}$; that is, at a much lower level. This does not mean that the neural events they could represent would be in the same ratio as the numerical values of Re/Im, nor does it mean that the behavioural importance of the $Y_{obs}(\text{Im})$ and $Y_{obs}(\text{Re})$ components is proportional exactly to 10^{-8}.

The $Mean[Y_{obs}(\text{Im})]$ values remain around the same average with increasing a, but they oscillate about their average more wildly. The asso-

64

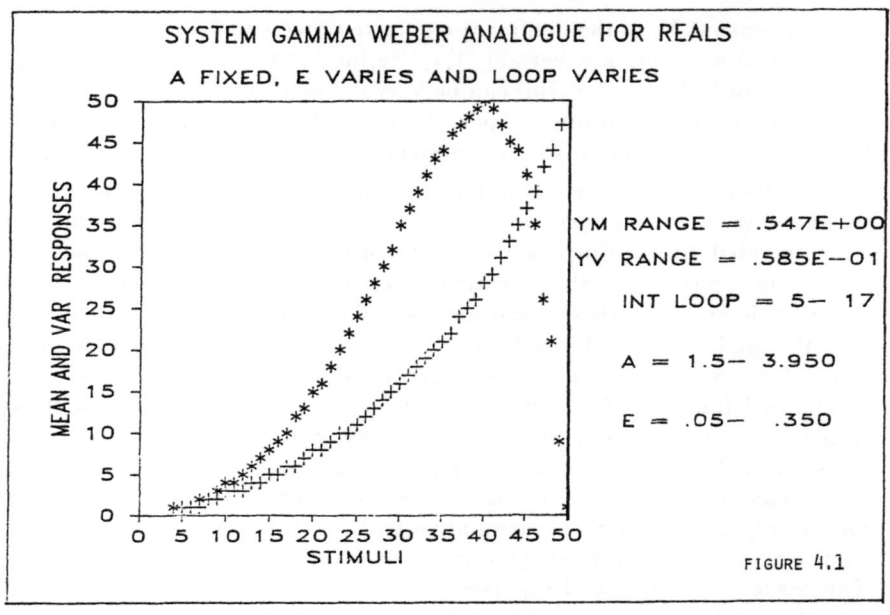

FIGURE 4.1

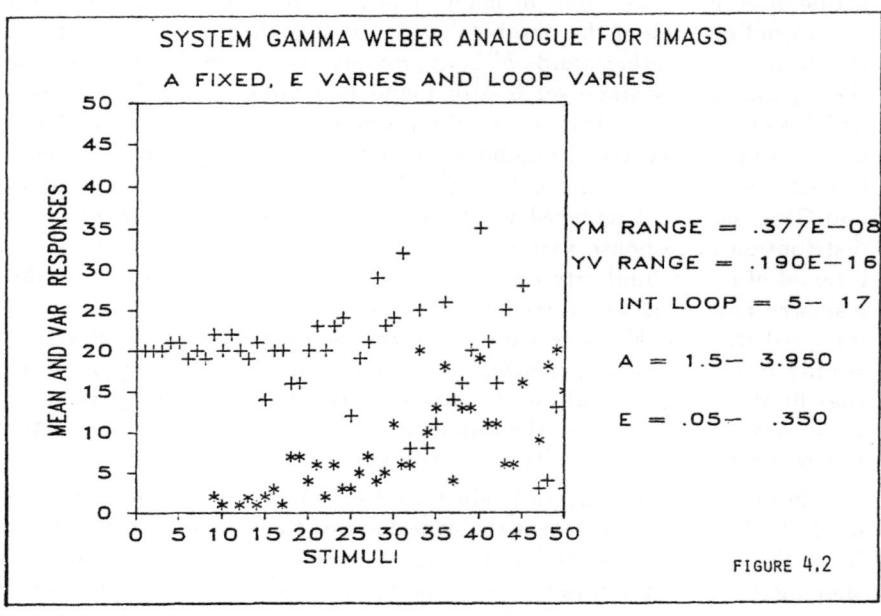

FIGURE 4.2

ciated variance $var[Y_{obs}(\text{Im})]$ increases and also shows progressively more scatter with increasing a; thus if we wish to use the moments of the distributions adequately to characterise them it would be necessary to go to the fourth powers of Y_η. It is clear that the chaotic noise in Γ is again revealed in the $Y_{obs}(\text{Im})$ component; that is to say, the partitioning of the system's output into apparently clean linear deterministic behaviour and residual noise is affected by the Re,Im mapping, and that noise increases with output magnitude. The fact that a system output from a Real input can be partitioned in this way has three implications: (i) it suggests why deterministic nonchaotic linear models with stochastic residuals will be a good first-approximation description of psychophysical input-output relationships, *if only the Re components are externally observable* , (ii) it indicates that the mathematical representation in complex form could be isomorphic with a split between externally observable system output and internal spontaneous system activity at a lower level (the neuropsychology corresponding to the psychophysics), and (iii) no necessary confirmation of this split is available from time series analyses solely of Re outputs (Gregson 1984, Gregson and Gates 1985).

The $var[Y_{obs}(\text{Re,Im})]$ for a given a is generated in Figures 4.1 and 4.2 by a bivariate lattice 7×7 of $e = .05$ to $.350$, $\eta = 5$ to 17. It is important to see if the relative contribution of e and η to the form of the $var[Y_{obs}(\text{Re})]/a$ function can be explored. Two other series of simulations are required as follows. Fixing η at values, in turn, of 2, 5, 10, and 20, and letting e vary over 49 values in the range $.008$ to $.392$ gives Figures 4.3 through 4.10. Considering first only Figures 4.3, 4.5, 4.7, and 4.9 for $Y_{obs}(\text{Re})$, the peak locus of the parabola increases wth η, the $\eta = 2$ case most closely resembles the Eisler *et al* (1977) data. The $\eta = 10$ and 20 cases are very skew, and the banana form is thus reproducible in this range. As the top of the input range would not be so readily observable in typical experimental data, the long almost straight line rise in $var[Y_{obs}(\text{Re})]$ corresponds to a nearly constant k' over a wide stimulus range combined with an open -ended response scale.

Why responding via a category scale should resemble the model with a low η and via an open-ended scale should resemble the model with high η is not clear, but if the resemblance is not fortuitous it leads to some testable predictions, because other properties of Γ arising solely from variations in η, both in the topology of the gain function and the autoregressive spctrum of the observable output, are also at the same time predictable. The model should therefore tie these things together. We have not at this time been able to trace any studies of the comparative autoregressive structure of response series to the same stimulus series, for comparable tasks with the only difference being the use of open-ended and closed-ended response scales

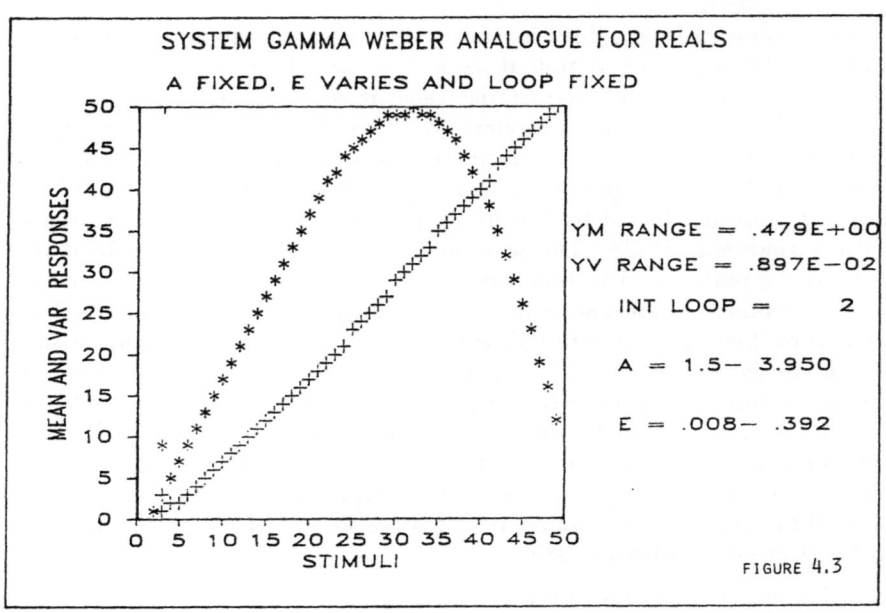

SYSTEM GAMMA WEBER ANALOGUE FOR REALS

A FIXED, E VARIES AND LOOP FIXED

YM RANGE = .479E+00

YV RANGE = .897E−02

INT LOOP =   2

A = 1.5− 3.950

E = .008− .392

FIGURE 4.3

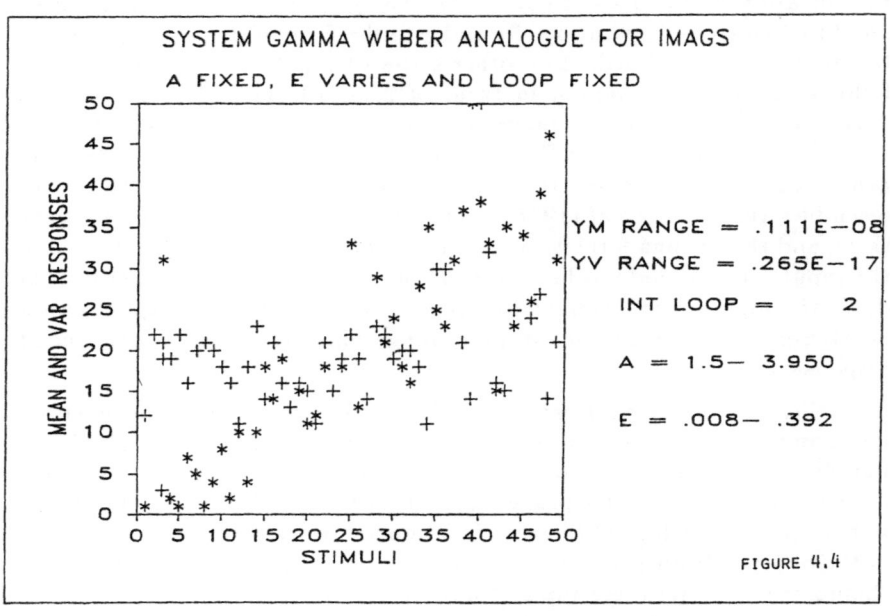

SYSTEM GAMMA WEBER ANALOGUE FOR IMAGS

A FIXED, E VARIES AND LOOP FIXED

YM RANGE = .111E−08

YV RANGE = .265E−17

INT LOOP =   2

A = 1.5− 3.950

E = .008− .392

FIGURE 4.4

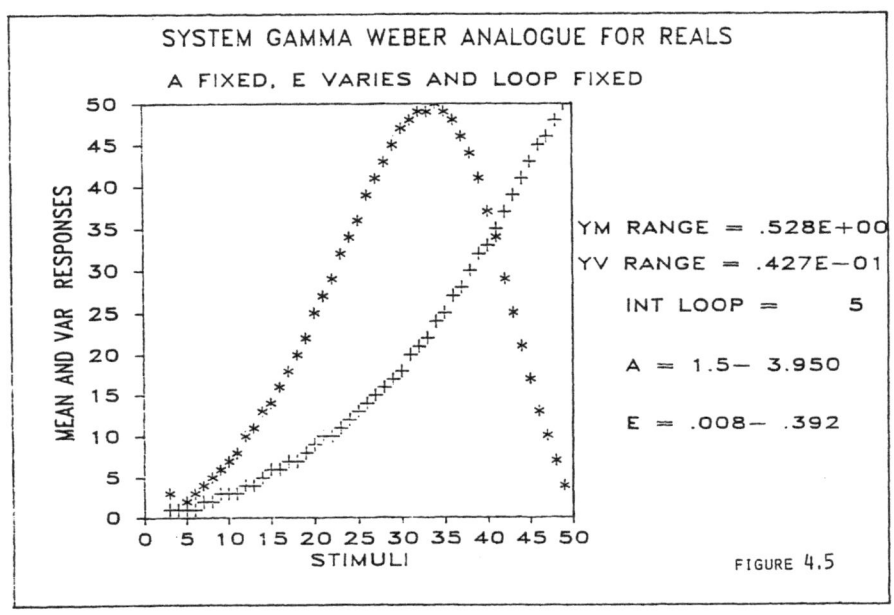

SYSTEM GAMMA WEBER ANALOGUE FOR REALS

A FIXED, E VARIES AND LOOP FIXED

YM RANGE = .528E+00

YV RANGE = .427E−01

INT LOOP = 5

A = 1.5− 3.950

E = .008− .392

FIGURE 4.5

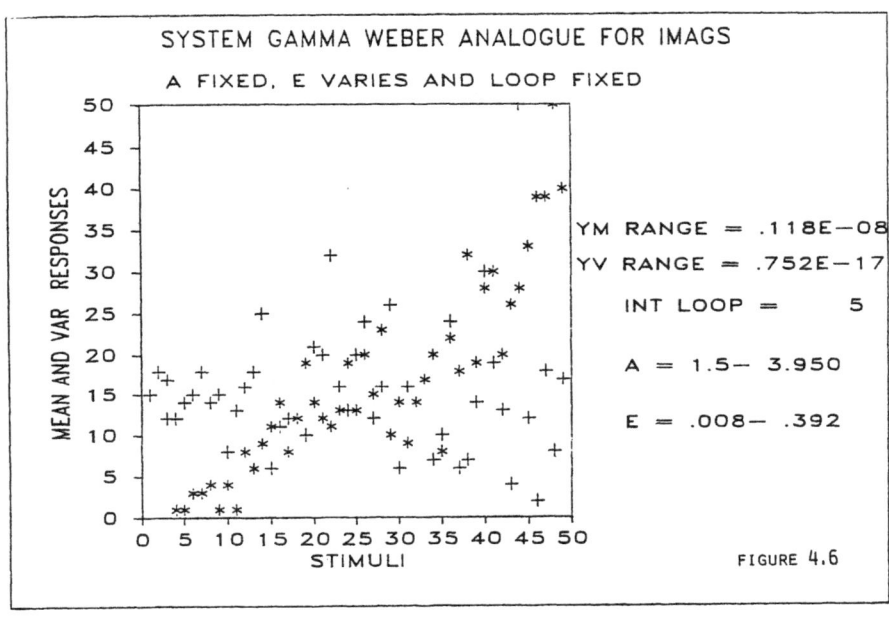

SYSTEM GAMMA WEBER ANALOGUE FOR IMAGS

A FIXED, E VARIES AND LOOP FIXED

YM RANGE = .118E−08

YV RANGE = .752E−17

INT LOOP = 5

A = 1.5− 3.950

E = .008− .392

FIGURE 4.6

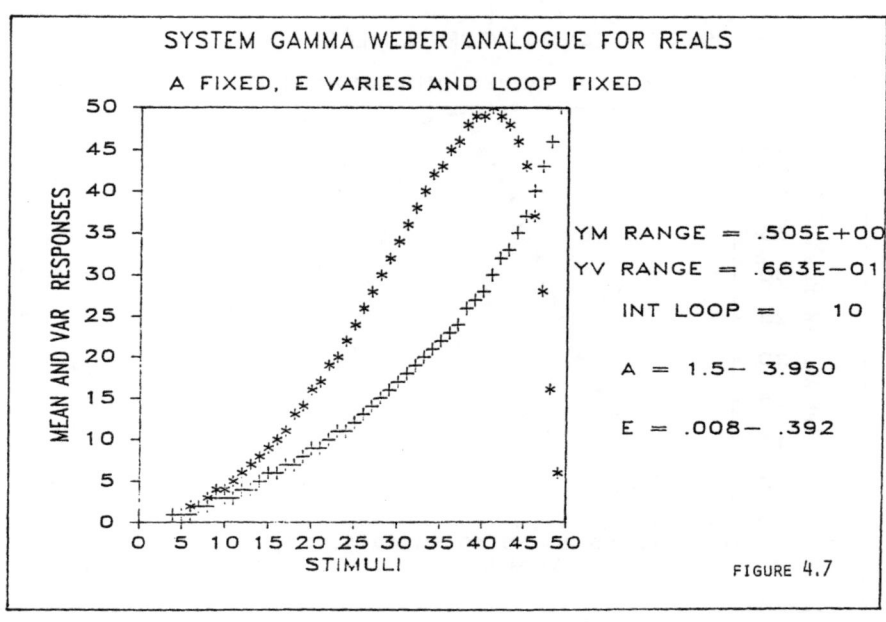

SYSTEM GAMMA WEBER ANALOGUE FOR REALS

A FIXED, E VARIES AND LOOP FIXED

YM RANGE = .505E+00

YV RANGE = .663E−01

INT LOOP = 10

A = 1.5− 3.950

E = .008− .392

FIGURE 4.7

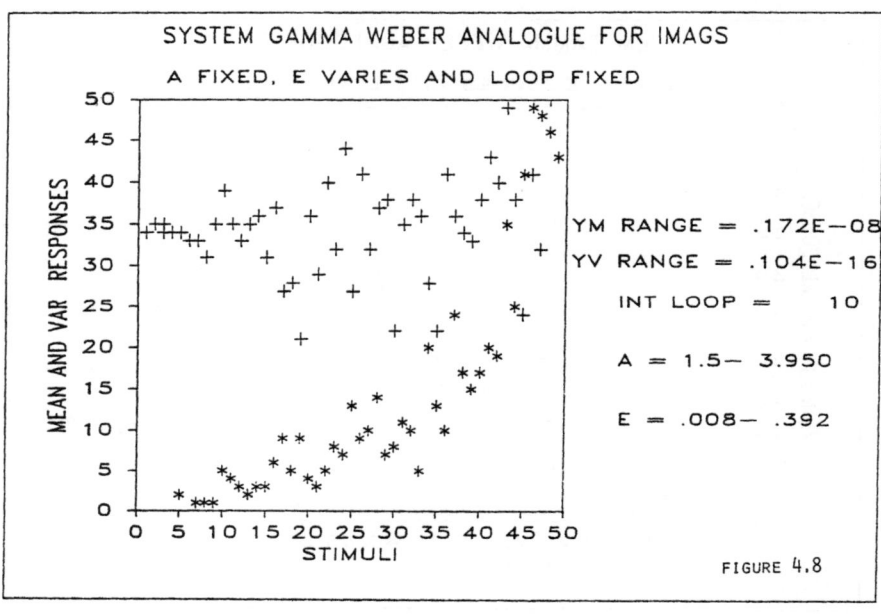

SYSTEM GAMMA WEBER ANALOGUE FOR IMAGS

A FIXED, E VARIES AND LOOP FIXED

YM RANGE = .172E−08

YV RANGE = .104E−16

INT LOOP = 10

A = 1.5− 3.950

E = .008− .392

FIGURE 4.8

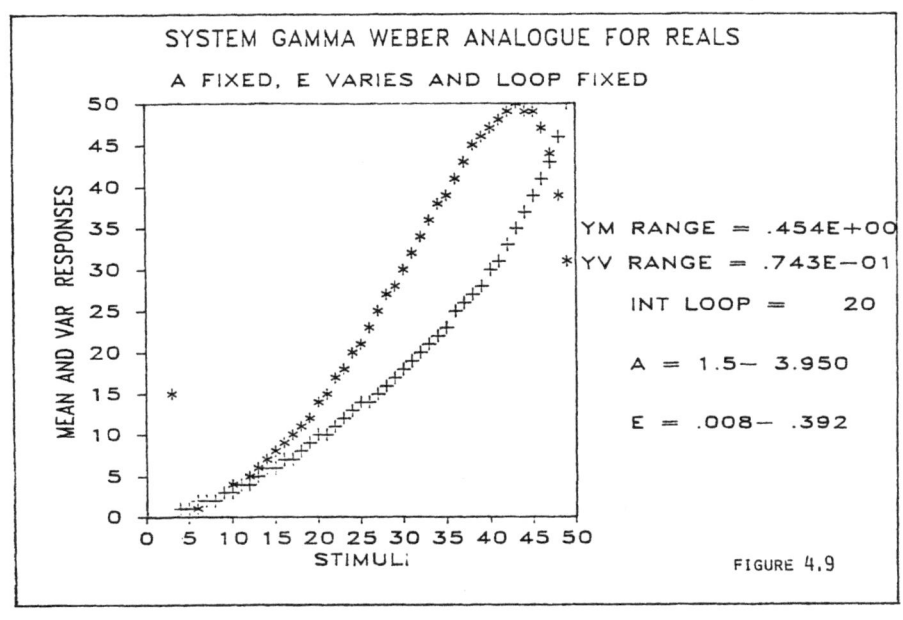

SYSTEM GAMMA WEBER ANALOGUE FOR REALS

A FIXED, E VARIES AND LOOP FIXED

YM RANGE = .454E+00

YV RANGE = .743E-01

INT LOOP = 20

A = 1.5- 3.950

E = .008- .392

FIGURE 4.9

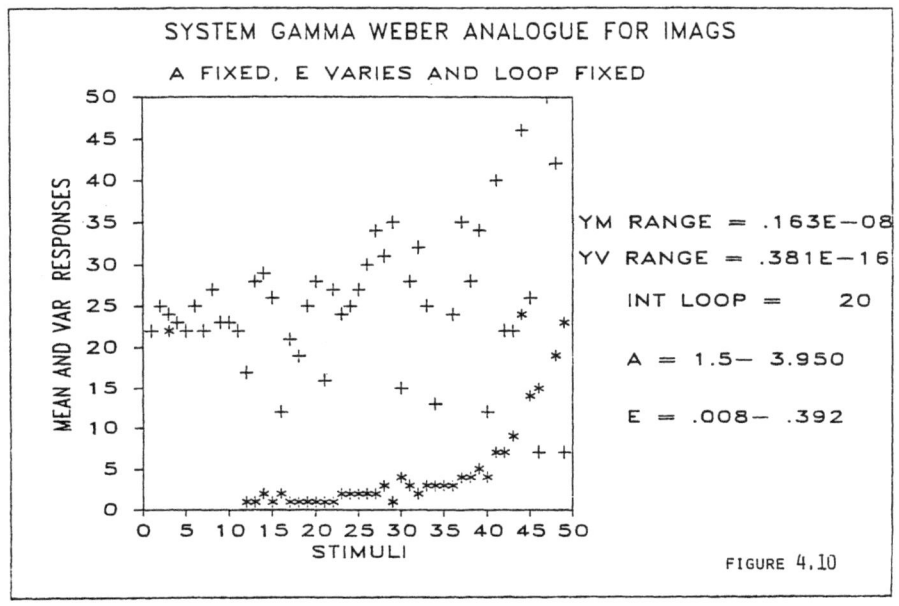

SYSTEM GAMMA WEBER ANALOGUE FOR IMAGS

A FIXED, E VARIES AND LOOP FIXED

YM RANGE = .163E-08

YV RANGE = .381E-16

INT LOOP = 20

A = 1.5- 3.950

E = .008- .392

FIGURE 4.10

imposed by the experimenter.

In contrast, fixing e and varying η by random integers over the range 5 to 25, again to give 49 values for each fixed a, η pair, is seen from Figures 4.11 through 4.16 to yield a qualitatively different picture. Now the $Mean[Y_{obs}(\text{Re})]$ rises in an ogival fashion, the degree of inflexion varying with e, but the associated $var[Y_{obs}(\text{Re})]$ has a very leptokurtic form with only a narrow range in which it is not negligible. An interim conclusion, therefore, is that varying e and η together for a fixed a provides a picture most like real reported data, so far as any data with sufficiently deep analysis are available for comparison. If η alone is fixed the picture is nearly as good, but fixing e and varying only η seems paradoxical in some respects.

Some reasons for the qualitative differences between Figures 4.3, 4.5, 4.7, 4.9 on the one hand, and 4.11, 4.13, and 4.15 on the other, may be elucidated by comparison of the topologies in the corresponding plots for $Y_{obs}(\text{Im})$; that is 4.4, 4.6, 4.8, 4.10 as against 4.12, 4.14, and 4.16. In the $Mean[Y_{obs}(\text{Im})]/a$ plots of Figures 4.4, 4.6, 4.8 , and 4.10, the fixed η variable e cases, there is shown a consistent pattern of increasing variance with a; the vertical floating between plots can be due to an outlier shifting the range of normalization and is not important. In all plots the $var[Y_{obs}(\text{Im})]$ increases with a; the trend becomes clearly discernable in Figure 4.10 which appears as a limiting case of the progressively diminishing scatter of plots 4.4, 4.6, and 4.8. In other words, the greater is η, the less is the uncertainty about the Im variance as a function of a. Obviously there is a relation between the locus of the peak of $var[Y_{obs}(\text{Re})]/a; \eta$ and the smoothness of the trend line of $v\bar{a}r[Y_{obs}(\text{Im})]/a; \eta$ but it cannot be said on this evidence that it follows from the intrinsic algebra in any formal sense.

In Figures 4.12, 4.14, 4.16 for fixed e and variable η the $Mean[Y_{obs}(\text{Im})]$ is constant (that is, independent of a), for all but the higher values of a. As e increases the point at which $Mean[Y_{obs}(\text{Im})]$ breaks into scatter moves lower in terms of a. The interplay between a and e in generating periodicity, bifurcation, and chaos first in the Im components, underlies this situation. Obviously the Y variance in the local $\{a, e, \eta\}$ parameter space depends upon some instability which is mainly absent over the lower range of a for which $Mean[Y_{obs}(\text{Im})] = const$. The role of e is dominant over that of η (as one might expect for larger η) in fixing both where the output variance lies on a parabola, and the leptokurticity (peakedness) of the parabola. It also appears to operate to convert an ogival form of $Mean[Y_{obs}(\text{Re})]/a$ into a negatively accelerated curve as in Figure 4.1.

The onset of oscillation and/or chaos in the $Mean[Y_{obs}(\text{Im})]$ appears to be a collateral phenomenon with the parabolic $var[Y_{obs}(\text{Re})]$ distribution. This suggests that the observable output variance is produced by the interaction of the internal variable e, which is strictly $(0, e)$ in Re,Im terms,

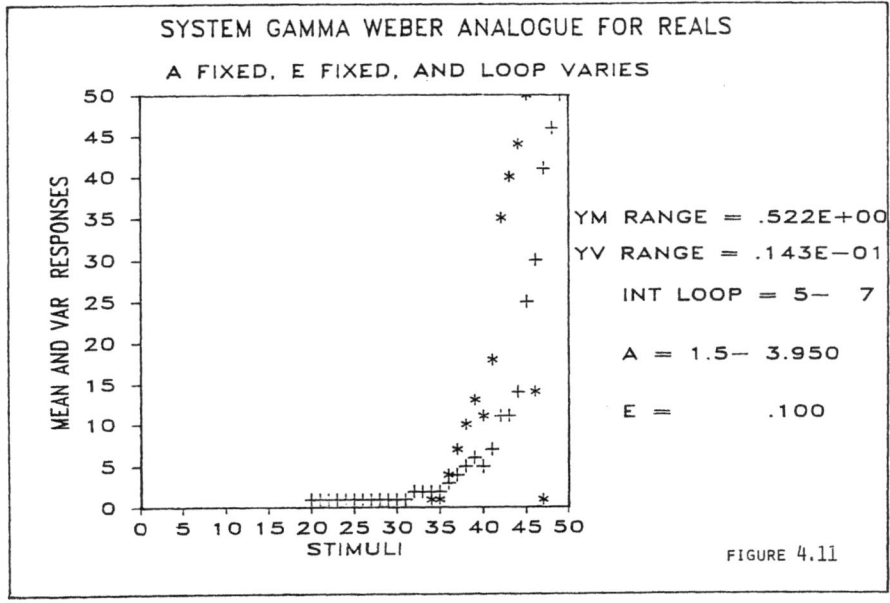

SYSTEM GAMMA WEBER ANALOGUE FOR REALS

A FIXED, E FIXED, AND LOOP VARIES

YM RANGE = .522E+00

YV RANGE = .143E−01

INT LOOP = 5− 7

A = 1.5− 3.950

E =   .100

FIGURE 4.11

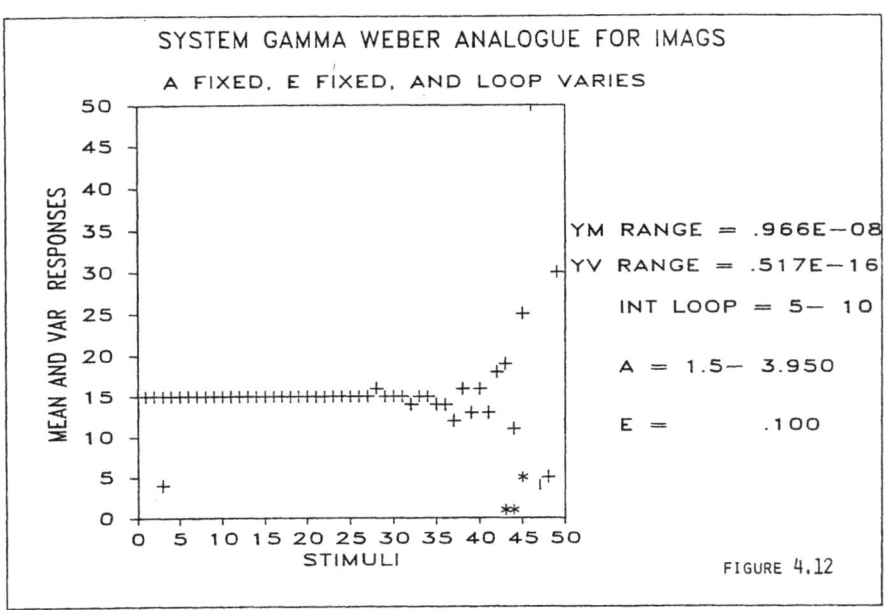

SYSTEM GAMMA WEBER ANALOGUE FOR IMAGS

A FIXED, E FIXED, AND LOOP VARIES

YM RANGE = .966E−08

YV RANGE = .517E−16

INT LOOP = 5− 10

A = 1.5− 3.950

E =   .100

FIGURE 4.12

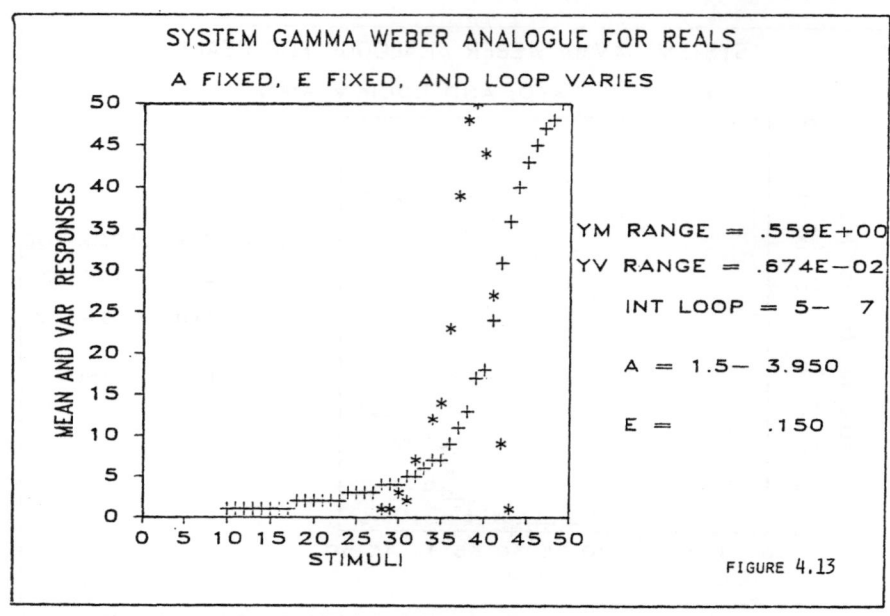

SYSTEM GAMMA WEBER ANALOGUE FOR REALS

A FIXED, E FIXED, AND LOOP VARIES

YM RANGE = .559E+00

YV RANGE = .674E−02

INT LOOP = 5− 7

A = 1.5− 3.950

E = .150

FIGURE 4.13

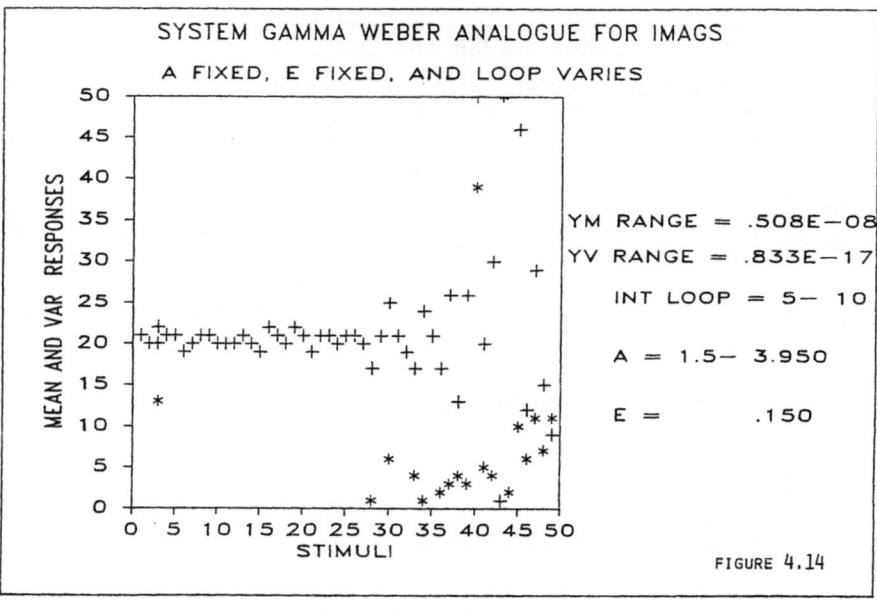

SYSTEM GAMMA WEBER ANALOGUE FOR IMAGS

A FIXED, E FIXED, AND LOOP VARIES

YM RANGE = .508E−08

YV RANGE = .833E−17

INT LOOP = 5− 10

A = 1.5− 3.950

E = .150

FIGURE 4.14

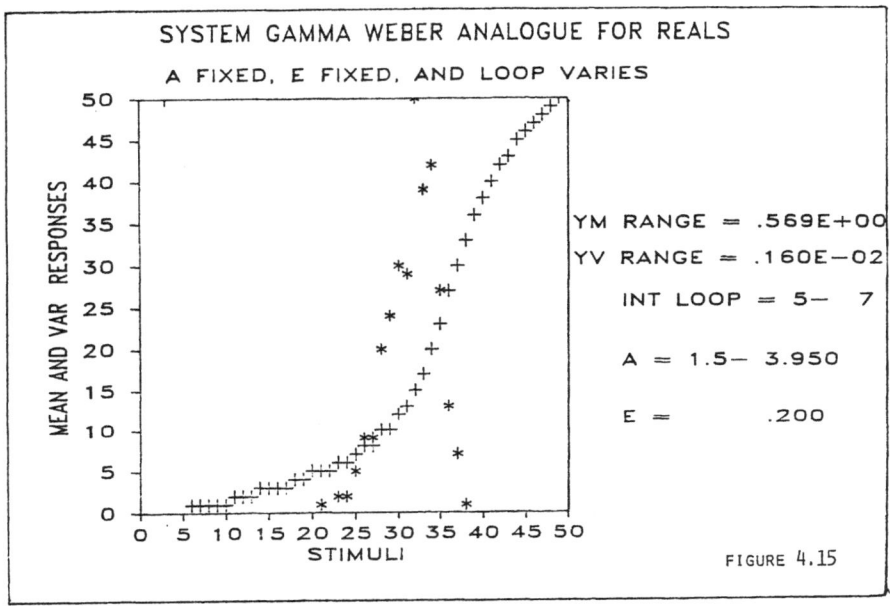

SYSTEM GAMMA WEBER ANALOGUE FOR REALS

A FIXED, E FIXED, AND LOOP VARIES

YM RANGE = .569E+00

YV RANGE = .160E−02

INT LOOP = 5− 7

A = 1.5− 3.950

E =   .200

FIGURE 4.15

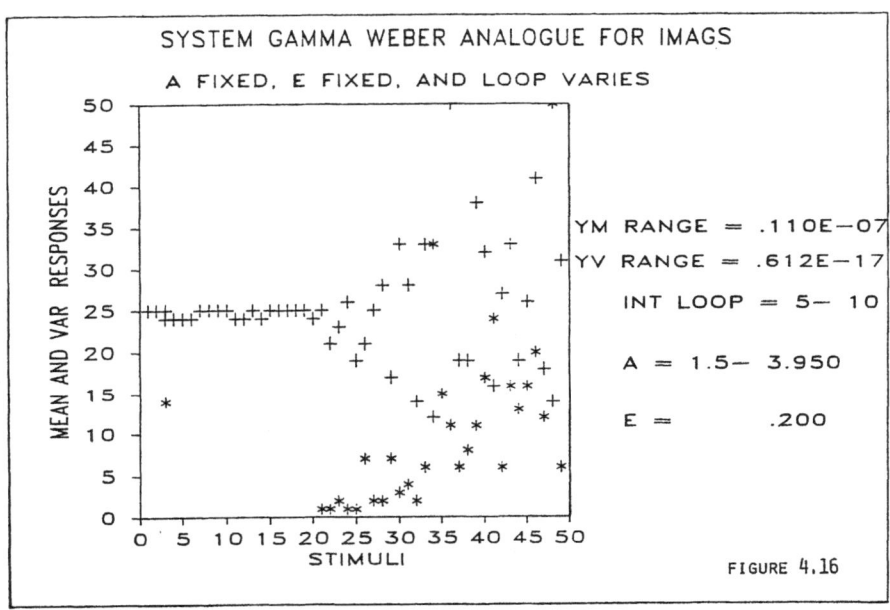

SYSTEM GAMMA WEBER ANALOGUE FOR IMAGS

A FIXED, E FIXED, AND LOOP VARIES

YM RANGE = .110E−07

YV RANGE = .612E−17

INT LOOP = 5− 10

A = 1.5− 3.950

E =   .200

FIGURE 4.16

with a which is Re. This could make sense in neurophysiological network terms; the uncertainty of information transmission through a network can obviously be modified by noise in that network particularly if it is correlated with the input. The noise here is deterministic: recently Eckmann, Thomas and Wittmer (1981) have shown that the intermittency properties of the difference recursion, which are here found in the Im components, are not destroyed by the injection of white noise in some cases. The observable result found here might thus persist even with superimposed stochastic noise from outside of Γ.

It should be noted that the upper value of INT LOOP in Figures 4.11 to 4.16 is not the maximum in the situation, but the last random value generated in a set of 49 points.

Modelling with Entropy Flow

The possibility of splitting the output from a nonlinear system such as Γ into three parts; (i) a deterministic (smooth) gain function, (ii) observable macrouncertainty, and (iii) internal microuncertainty, and also varying the topology of (i) and the relative role of (ii) and (iii) as a function of the system parameters has been shown to be possible. It is a potentially rich and complicated situation, but with sometimes quite regular observable input-output relations. In this particular analysis the split between Re and Im apparently matches that between macro- and microinformation flow, the microinformation in turn being interpreted as physiological rather than psychological.

This is not, of course, the usual way to write a model in psychophysics. To make the situation more formal, consider the contrast between

$$Y = U + \mathbf{N}(\bar{\xi}, \sigma_\xi) \qquad \xi \text{ is noise} \qquad [4.1]$$
$$\text{and } Y = U + \mathbf{D}(\bar{\chi}, \sigma_\chi) \quad + \rightleftharpoons \quad \Gamma(\omega; a, e, \eta). \qquad [4.2]$$

In [4.1] the observables are the response, Y, whose observable noise variance σ_y^2 is determined by its covariance with the normally distributed variable ξ. This is the 'classical' way of writing much of psychometrics and regression theory. As soon as we consider Γ then the formal representation is like [4.2], χ is the macro noise variance, with its own variance σ_χ associated with some distribution $\mathbf{D}$ (of which the parabolic form for the subjective Weber function may be an example), and where $\mathbf{D}$ may also have U as a parameter, and where the internal noise ω is known to be a function of $a, e,$ and η.

But [4.2] does not properly capture what is happening in a nonlinear system (the form of [4.2] serves simply to heighten the contrast with the assumptions in [4.1]) because the strict division between U and ξ in [4.1] is not matched by a strict division between each of the terms in [4.2]. Rather, the characteristic of nonlinear difference systems is a possibility for entropy flow between the deterministic and the uncertainty phases in the system (Shaw, 1981). The turbulence in the Γ microuncertainty flows into the χ and eventually into the U phases of the system, thus making it unpredictable. We have attempted to emphasise the dynamic uncertainty in the gross system description (by contrast local behaviour in a very restricted U region corresponds to a small a range and is thus possibly well-behaved) by introducing the symbol "$+ \rightleftharpoons$" in [4.2].

References

Eckmann, J.- P., Thomas, L., and Wittmer, P. (1981) Intermittency in the Presence of Noise. *Journal of Physics, 14A*, 3153 - 3168.

Eisler, H., Holm, S., and Montgomery, H. (1979) The General Psychophysical Differential Equation; a Comparison of three Specifications. *Journal of Mathematical Psychology , 20*, 16 - 34.

Ekman, G., and Künnapas, T. (1969) Distribution Function for Similarity Estimates. *Perceptual and Motor Skills, 29*. 967 - 983.

Gregson, R. A. M. (1975) *Psychometrics of Similarity*. New York: Academic Press.

Gregson, R. A. M. (1976) A Comparative Evaluation of Seven Similarity Models. *British Journal of Mathematical and Statistical Psychology, 29*, 139 - 156.

Gregson, R. A. M. (1984) Behaviour of a system with gain and pure delay filters incorporating a nonlinear difference feedback loop as a generalized psychophysical model. *Unpublished Seminars, Federal Republic of Germany*.

Montgomery, H., and Eisler, H. (1974) Is an equal interval scale an equal discriminability scale? *Perception and Psychophysics, 15*, 441 - 448.

Shaw, R. (1981) Strange Attractors, Chaotic Behavior, and Information Flow. *Zeitschrift für Naturforschung, 36a*, 80 - 112.

5 A Generic Dynamic Mapping of Environment onto the System

So far we have considered what is perhaps the simplest viable form of Γ within a system, where only $U \Longrightarrow a$, but the remaining parameters were either (i) fixed, or (ii) varied randomly in time at a second-order level, or (iii) could be externally varied nonrandomly but independently of U. Case V 7 was not devised first (Gregson, 1984) but was considered here first because of its relatively direct relation to the CNO-generating property which is of interest in psychophysics. However, though V 7 is well-behaved in some ways, for that very reason it cannot be used to mimic some of the more ill-behaved phenomena listed in A1 to A19 of Chapter 2. Also, the V 1 to be considered here is the starting point of a number of other variants which are readily produced by relaxing or removing constraints on the parameter bounds and on the maximal range of rates of change from one cycle to the next. The idea now is to set up the most complicated identifiable structure that might be plausibly studied, and then to consider it and a number of cases derived by simplifying or constraining the general case. This is not simply a mathematical device, it also suggests ways in which transient lesions in mapping between parts of a dynamic system can readily induce the sort of phenomena which are reported in experiments where there is a failure to obtain a monotonically increasing $S - \mathcal{R}$ psychometric function like a CNO.

Consider Figure 5.1. This is a general framework from which Γ V 7 can be derived as a special case. The number of possible mappings W is as wide as the unfettered imagination of a theorist can devise, but only a few are worth consideration here, to show

(i) the great quantitative and qualitative changes in the input-output topology function of Γ which arise from extremely small parameter changes or boundary constraints,

(ii) the robustness of the system with respect to other parameter changes,

(iii) the capacity of the system to mimic a diversity of psychophysical phenomena,

(iv) the occurrence of some pathological cases in neighbourhoods of the parameter space $\{a, e, \eta\}$ which resemble behaviour arising in neurological disorders,

(v) that displacement from a current input-output topology by change to another *adjacent* region of the parameter space transforms into a topology which is induced experimentally in real data by a change in the conditions corresponding to the change in the parameter space.

This last condition (v) has been considered by Smale (1980) as it raises problems in defining the distance between two dynamic systems. The idea here is that a nonlinear system can show great changes in its input-output relations due to very small changes in its internal parameters, in regions where it is not parameter robust. But when such changes are induced they are lawful in the sense that one dynamic system associated with a subregion of the Γ system parameter space can only transfer with small parameter changes (into an adjacent subregion) into another dynamic system which is associated with the adjacent subregion. If the model is applicable to real data, then small changes in W cannot induce changes in the input-output topologies which are associated with a shift from one parameter space subregion to another remote subregion in $\{a, e, \eta\}$.

The conditions imposed on the Γ generic structure of Figure 5.1 for V 1 are

$$U_J \sim \text{Rect}(z), \qquad J = 1, \ldots, 50 \qquad\qquad [5.1.1]$$
$$\text{where } z \text{ is random, and}$$
$$\text{Rect is a rectangular probability distribution on } (0, 1)$$
$$U_{min} \leq U_J \leq 3.99 \qquad\qquad [5.1.2]$$
$$a_J = U_J \quad \text{in } [5.1.2] \qquad\qquad [5.1.3]$$
$$\Delta^1 a_J \leq e_{max} \qquad\qquad [5.1.4]$$
$$ie_J = i\Delta^1 a_J \qquad\qquad [5.1.5]$$
$$Y_0 = 0.5, \epsilon \quad \text{(complex)} \qquad\qquad [5.1.6]$$
$$\eta = 5 \qquad\qquad [5.1.7]$$
$$e_{max} = \kappa = const, \quad 0 < \kappa < 1 \qquad\qquad [5.1.8]$$

The point of using [5.1.1] is that a random demand characteristic on the system resembles the practice in designing psychophysical experiments. It

System   Γ'

Coupling of the Environment to
the internal feedback loop

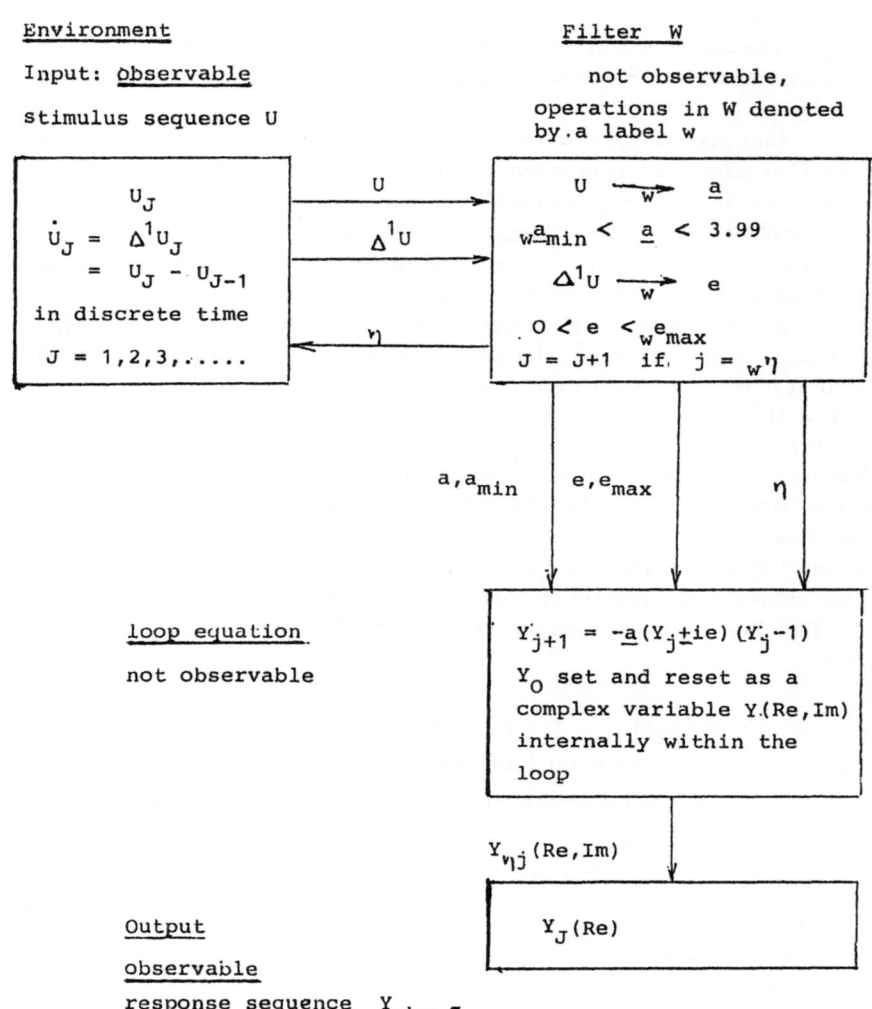

Environment

Input: <u>observable</u>

stimulus sequence U

$$\dot{U}_J = \Delta^1 U_J$$
$$= U_J - U_{J-1}$$
in discrete time
$$J = 1,2,3,\ldots$$

<u>Filter</u>   W

not observable,
operations in W denoted
by a label w

$$U \xrightarrow{w} \underline{a}$$
$$w\underline{a}_{min} < \underline{a} < 3.99$$
$$\Delta^1 U \xrightarrow{w} e$$
$$0 < e <_w e_{max}$$
$$J = J+1 \quad \text{if} \quad j = {}_w\eta$$

a, a_{min} e, e_{max} η

<u>loop equation</u>

not observable

$$Y_{j+1} = -\underline{a}(Y_j \pm ie)(Y_j - 1)$$
Y_0 set and reset as a
complex variable Y.(Re,Im)
internally within the
loop

$Y_{\eta j}$ (Re,Im)

<u>Output</u>

<u>observable</u>

response sequence $Y_{obs,J}$

Y_J (Re)

Figure 5.1

does not resemble the real world in general. The condition [5.1.8] is effectively an additional system parameter.

Simulations are run over a region of the bivariate space of U_{min}, ϵ_{max}, within the limits

$$1.8 \leq U_{min} \leq 3.2, \qquad .05 \leq \epsilon_{max} \leq .25$$

and some selected results are shown in Figures 5.2 through 5.5.

One consequence of [5.1.4], because $max \mid \Delta^1 a \mid = 1$ and $\epsilon_{max} < 1$, is that the distribution of $\Delta^1 a_J$ has a lump of probability at the upper limit, whereas [5.1.1] does not, because of [5.1.4]. It is thus not true that $\Delta^1 a_J \sim Rect(z)$. Of course, even without [5.1.4] the distribution of successive $\Delta^1 a$ for $a = Rect(z)$ is (Johnson and Kotz, 1970) $E(\Delta^1 a) = 0$, $var(\Delta^1 a) = 2.var(a) = .167$, and $\beta_2(\Delta^1 a) = 1.8$, which is not uniform over $(0,1)$ for $\mid \Delta^1 a \mid$. The lump of probability induces an uneven density in the $U \Longrightarrow Y_{obs}$ plot, given a [5.1.1] which has a uniform density in U.

For simulations $\eta = 5$ was chosen after a family of test runs over the range $\eta = 2, 5, 10, 20, 50, 200$ [1] showed in most cases an input-output plot tending to a straight line for $\eta = 5$ or 10. We are thus exploring system properties in the preasymptotic dynamics of Γ. It makes but little difference, if $\eta > 10$, what is used in this sense.

Figures 5.2 through 5.5 are like later graphs recurring in this monograph of similar form; the computer graphics used the convention that + is a single data point, ∗ is two coincident points, ⋆ is 3 or 4 coincident points, and a six-pointed star implies 5 or more coincident points.

The effect of increasing ϵ_{max} is to allow a second breakaway CNO-like distribution to appear below the main ogive. The interest in systems which appear to have two $S - R$ functions and to jump between them, arises in various areas of psychophysics, particularly in time perception, and is briefly reviewed by Gregson (1983, sect.4.3). There are various reasons why a subject might apparently jump between two states, each associated with one CNO-like function, and different transfer functions have been in some models tentatively associated with different ranges of stimulus intensities, which here would map onto different ranges of a. The point here is to show that a double plot like Figure 5.4, say, which is not strictly two continuous functions but two distributions each locally dense in a neighbourhood which resembles a fuzzy line, can emerge if the system parameters are sensitive, with an upper dynamic boundary, to rates of change of inputs as well as to inputs.

---

[1] Conducted at the Institut für Psychologie, Technische Universität Braunschweig in 1984 on a PDP11/34, and in part replicated in the Psychology Laboratories of the University of New England in 1985 on a DEC20 computer.

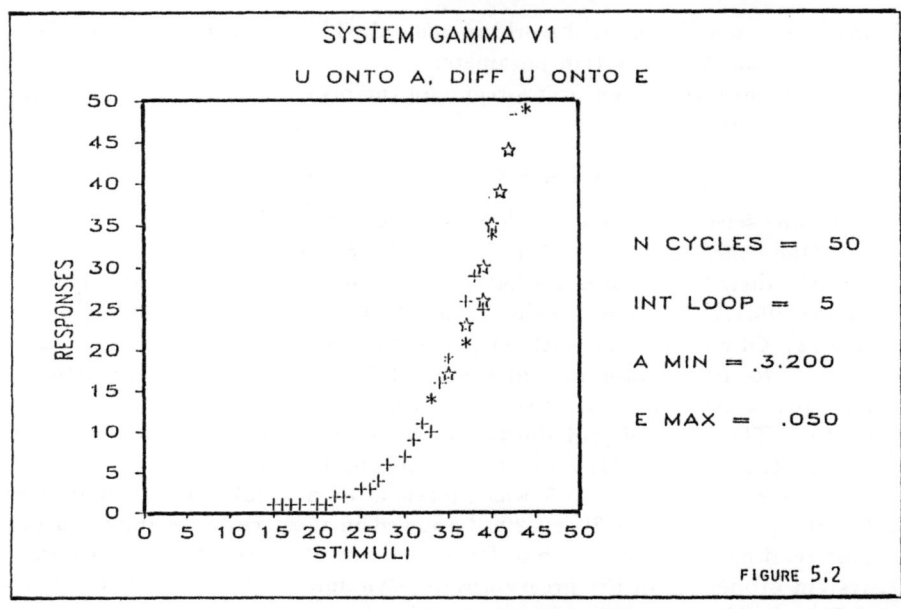

FIGURE 5.2

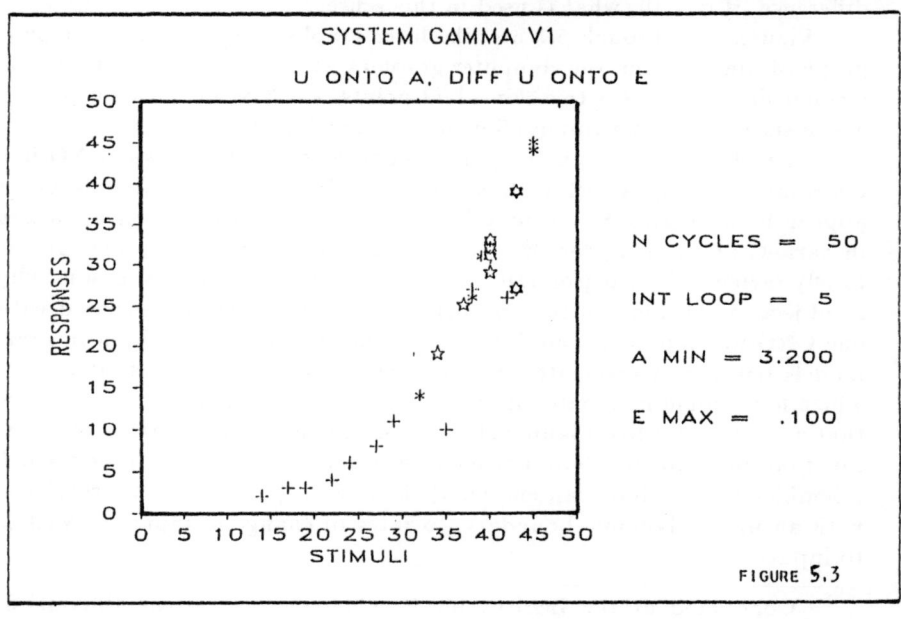

FIGURE 5.3

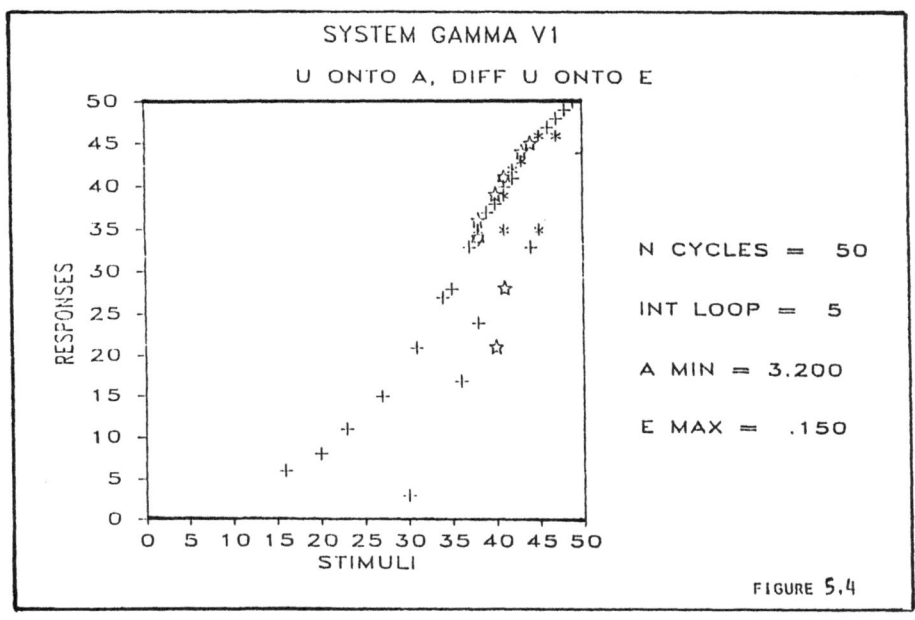

FIGURE 5.4

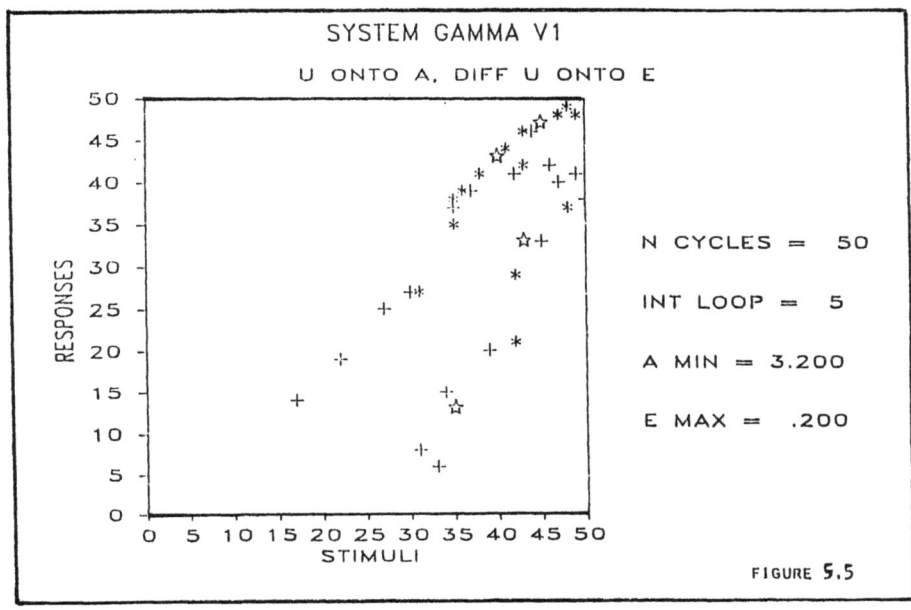

FIGURE 5.5

The case Γ V 1 corresponds to an organism imposing its own bounds on admissible $\partial U/\partial J$ but none on U. In reality one might expect bounds, as a consequence of the limited psychophysical transduction of receptor mechanisms and neural pathways to the brain, on both. If ζ is large then the effect of ζ is negligible, except that it might then be possible for the system to encounter local combinations of ae which would induce a transient explosion if η is also large.[2] As ζ increases its effect decreases because most of the e values are then less than ζ; what then happens is that the system jumps, from trial to trial, (J to $J + 1$) around the two-dimensional parameter space $\{a, e \mid W, \eta\}$. This ideas leads us into the possibility of defining an experiment, which is formally a series

$$\{U\} \quad \Longrightarrow \quad \Gamma(w, a, e, \eta) \quad \Longrightarrow \quad \{Y_{obs}\}$$

as a constrained pathway through $\{a, e\}$. If e is fixed independently of the input, as in V 7, the pathways are on the a_{const} contours in the $\{a, e\}$ space of Table 2.1, and yield a variety of CNO-like forms approximated by cubic polynomials in $Y_{obs} = f(a, e \mid \eta)$. If we permit or induce, as here, some restrained jumps in e_J coupled with current $\Delta^1 a_J$, then the system "dances" between different functions; the relative time which it spends in different regions of the input-output topology depends on the statistical properties of the distribution of U as a time series. In this sense a system which is intrinsically completely deterministic will exhibit stochastic properties which are induced both by the environment and also (in the Re output) quite separately due to the nonlinear internal dynamics when ae is sufficiently large.

Identification to within local regions of the parameter space

The input of a random sequence U to Γ V 1 does not produce a corespondingly random sequence Y_{obs}, the parameters $\{a, e, \eta\}$ all affect to some extent the output autoregression spectrum, and the system is particularly sensitive to the action of e in this respect.

By simulation over a range of $a_{min} = 2.6$ (0.2) 3.4 and $e_{max} = .05$ (.10) .35. for $\eta = 5$, $J = 1, 2, \ldots, 100$ it appears that the coefficients of a pure autoregressive model in Y_{obs} show a form which resembles some results commonly found in real data (for examples see Gregson, 1983). In this simulation the variances of inputs and outputs have been made

---

[2] This circumstance is reminiscent of some theories about *schizophrenia* in which it is postulated that the maximum rate of information handling in the patient is very limited, and breakdown occurs if it is overloaded. This is purely an interesting speculation here.

arbitrary by linear transformations, consequently the AR coefficients are used and not the corresponding transfer function coefficients. The pattern of relative values and signs remains substantially unchanged.

Linear transfer coefficients commonly show a pattern such that a high initial positive term dominates, suggesting why an open-loop gain function is sometimes a reasonable approximation, and the subsequent damped oscillations are compatible with a second-order feedback expression. In the factorization of such results a negative real root and one negative conjugate pair usually dominate, both being outside the unit circle.

If, then, Γ is to simulate realistic psychophysical processes, it should have outputs which are in fact significantly autoregressive, and the coefficients should die away, with oscillations, in an exponential manner. This result only holds for a majority of human subjects in relatively easy tasks; a minority always exhibit more complex dynamic behaviour, or go into limit cycles, which can sometimes be identified by complex roots in the righthand halfplane (Schuster, 1984, gives an example).

As the variation in the AR coefficients over the range AR(1) to AR(7) is relatively minor with respect to changes in a_{min}, yet with e_{max} quite marked for the lower k of AR(k), we have averaged some results over the a parameter space explored for fixed e, and show these diagrammatically. It is seen that the pattern found in data (as for example reported by Michon, 1967 in time estimation experiments) is best approximated by low values of e_{max} and that above $e_{max} = .15$ the system is relatively insensitive to parameter changes. That is, the behaviour most similar to real data enables us to restrict plausible values of e_{max} to a small part of the parameter space, where it is also the case that small changes in e_{max} produce discernable changes in the output, for fixed a and η. In short, the system is not parameter robust with respect to e in the region where it most resembles real data. As a first approximation, therefore, $e_{min} = .07$ is a suitable point about which to explore the parameter space, given $2.6 < a_{min} < 3.4$, in order to maximise agreement between observed and theoretical AR(k) values. All the solutions to AR(1,...,7) used here have negligible residual autoregressions. It was decided to stop with 7 lags, as further estimates of parameter values became correlated beyond that point, and secondly because human response processes are not very likely to persist much further (though some apparent counter-examples have been found, see Gregson, 1986). The fitted coefficients are in most cases negligible for more than five lags, but sometimes a delayed peak at AR(7) is found with high e values. See Figure 5.6.

The relation between e_{max} and AR(k) is explicable. As e_{max} decreases, the forced AR(1) component increases. As e_{max} increases the constraints on $\Delta^1 Y$ are relaxed and the process approaches a random series with all

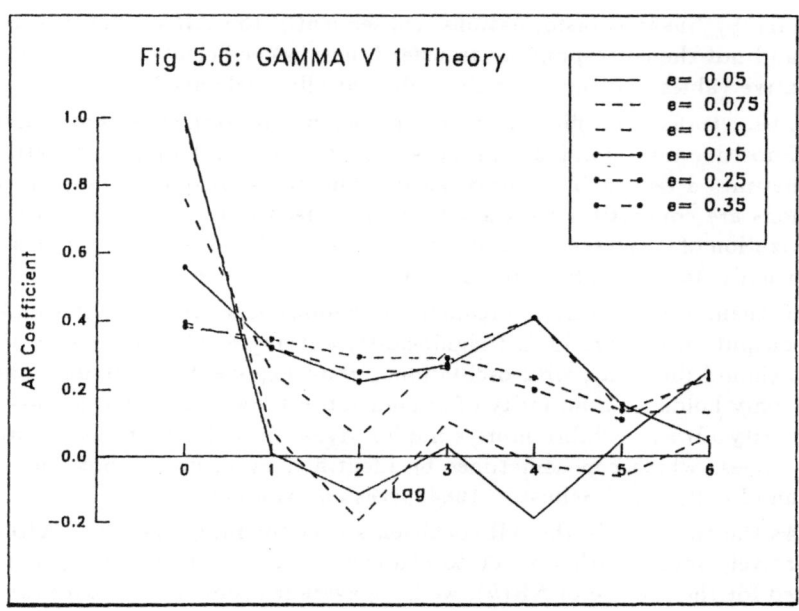

Fig 5.6: GAMMA V 1 Theory

AR(k) tending to equality.

The internal dynamics of the inner loop of Γ are obviously nonlinear and must vary greatly with a and with e, but interestingly the observable output of the sampled loop at $\eta = 5$ can be well approximated by a linear AR process of the form reported, providing that no cascading takes place, as will be examined later.

The compatibility of the Γ V 1 model with a real set of data might thus be assessed by

(1) Assuming that e_{max} is around .07, or a value near which is compatible with the observed AR(k) structure,

(2) Fixing a_{min}, given e_{max}, to produce a gain function topology resembing a psychophysical CNO function,

(3) Seeing if the response distribution entropy $\mathbf{H}(Y_{obs})$ is then degraded to some extent as in the model with the parameters estimated from (1) and (2).

If all three criteria can be met, then the locus of the data in the parameter space for Γ V 1 is limited to a subregion in which there is internal parameter robustness. The variability of η is a second-order tuning facility to improve fit marginally, but should not *a priori* be moved away from $\eta = 5$ if V 1 is a valid choice from the range of Γ variants considered. If the data exhibit any pathological forms other than the branching shown in Figures 5.2 to 5.5, such as breaks, then V 1 is not appropriate. Satisfying unusual topologies of the input-output relations should be a prime consideration, which does limit the choice of variant considerably.

Table 5.1

Mean l.t.f. Coefficients over the range $2.6 < a_{min} < 3.4$
with $\eta = 5$, for $\text{Rect}(U)$, $0 < U < 1$

e_{max}	.05	.075	.10	.15	.25	.35
v_0	1.005	.976	.757	.556	.380	.393
v_1	.009	.073	.250	.318	.344	.320
v_2	-.115	-.195	.059	.222	.292	.249
v_3	.030	.105	.312	.272	.290	.262
v_4	-.192	-.024	.404	.406	.232	.197
v_5	.050	-.061	.135	.154	.137	.109
v_6	.255	.061	.130	.044	.229	.248

see Figure 5.6 and Michon (1967)

References

Gregson, R. A. M. (1983) *Time Series in Psychology.* Hillsdale, New Jersey: L. Erlbaum Associates.

Michon, J. A. (1967) *Timing in Temporal Tracking.* Soesterberg: Netherlands Institute for Perception RVO-TNO.

Schuster, H. G. (1984) *Deterministic Chaos.* Weinheim: Physik-Verlag.

Smale, S. (1980) *The Mathematics of Time.* New York: Springer-Verlag.

6   Further Variants on Mappings of Inputs

Variations in the way that the input series, U, and its first difference $\Delta^1 U$, are coupled into the recursive loop parameters, as summarised in the filter W in Figure 5.1, affect profoundly the observable $U \rightarrow Y_{obs}(\text{Re})$, or stimulus $\rightarrow$ response relationships. As the topology of these relations is the prime interest here, and the numerical values mean little, the graphs produced have, for pictorial clarity, quite arbitrarily been scaled at 0 - 50 for the stimuli (actually $0 < U < 1$), and 0 - 50 for the responses, which given the U range and the $\{W, a, e, \eta, \zeta, \epsilon\}$ sets used can range over a great diversity of numerical values. This process of rescaling for the graphs has been referred to as *normalization*. In some cases the actual values are shown in the output of the computer graphics program employed.

The interest here is in seeing if plausible, or implausible (even absurdly impossible) results can be created. If all the results for all mappings W had the same topology, then the model Γ would be useless. If the creation of topologies which it is known can really occur were to be associated only with extreme (or for a worse example, negative) internal parameter values, then this of itself is sufficient to reject a model. Counter-intuitive results are admissible, but only within what seem reasonable bounds at the time.

Case V 2

If the input U is disconnected from a, and a is instead a fixed constant value, but at the same time e remains coupled to $\Delta^1 U$, with a multiplying factor, then the input induces a sort of responding to rates of environmental change, but not to actual stimulus levels. This case is interesting to consider, because some theorists (Helson, 1947, Krantz, 1972) have argued that responding is only made to change, and never to static levels of stimulation. If stimulus levels remain fixed for a long time then the observer may cease to be aware of any stimulation; adaptation is complete. The chemical senses, and touch, may show such phenomena.

The results of setting

$$\textbf{V2Ax1}: \quad U_{J+1} = U_J + rect(z - z_0)$$
$$\textbf{V2Ax2}: \quad a_J = a_{min} = const$$
$$\textbf{V2Ax3}: \quad ie_J = \Delta^1 U_J \cdot W_e,$$

where W_e is a scalar multiplier, are startlingly different from those of Γ V 1. Effectively responses are bimodal on a stimulus range z. This represents a serious degradation of information; the plot of stimulus - response values resembles the distribution of confidence or subjective probability in a task where only extreme responses are induced, for example in a relatively easy "same-or-different" discrimination problem. It also resembles a situation which occurs sometimes in the senses of taste or smell (Gregson, 1986), where an observer can only detect the presence of a stimulus but not assess its intensity.

Case V 3

This case is like V 2, but the starting value $Y_{0,J}$ is not fixed for each recursion at $(0.5, \epsilon)$, but is instead the last value of the previous iteration; $Y_{\eta, J-1} = Y_{0,J}$, $J > 1$. This makes very little difference and is therefore not shown pictorially. However, the distinction between Γ V 2 and Γ V 3 is neuropsychologically meaningful.

Case V 4

A consequence of restricting $\Delta^1 a_J < C$ is that the distribution of a values has a lump of probability at its upper limit; all stimuli above some value are equipotential, as in an upper threshold of perceived intensity. Deleting this constraint, but retaining

$$\textbf{V4Ax1}: \quad U_{J+1} = U_J + rect(z - z_0) \cdot W_a, \quad W_a < 1$$

where W_a is a scalar gain operator, and retaining

$$\textbf{V4Ax2}: \quad U_{min} \leq U_J \leq 3.99$$
$$\textbf{V4Ax3}: \quad a_J = U_J + a_{min}$$
$$\textbf{V4Ax4}: \quad ie_J = \Delta^1 a_J$$

the lump of probability on $\Delta^1 a_J$ is removed, and the case is intended now to correspond to an environmental limitation on $\Delta^1 U$ (like choosing to

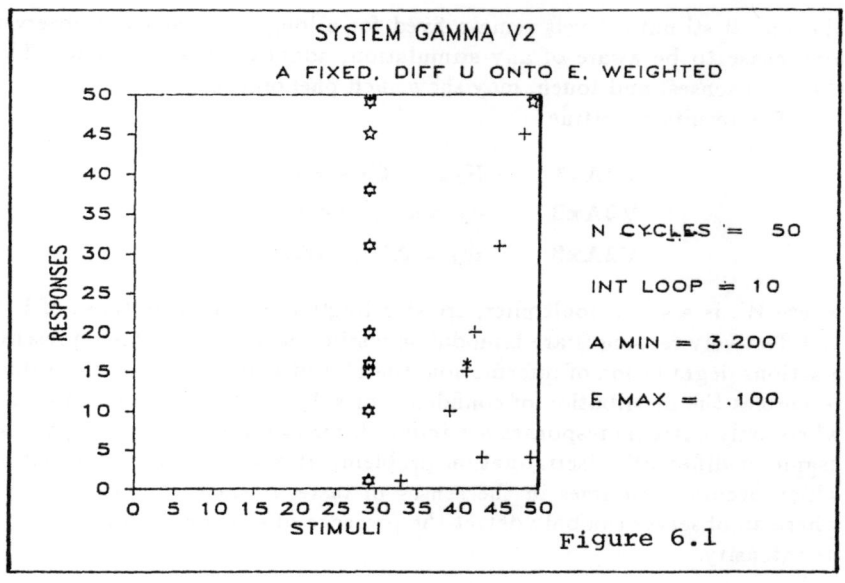

Figure 6.1

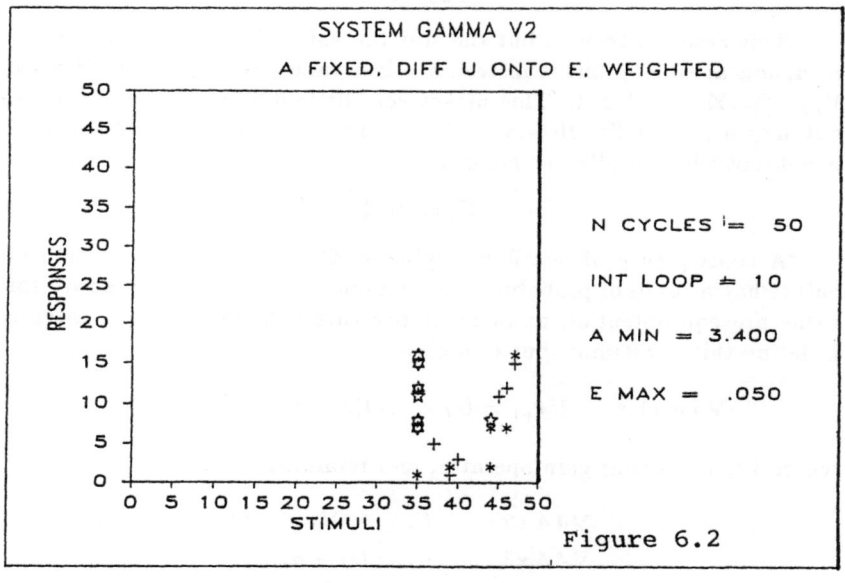

Figure 6.2

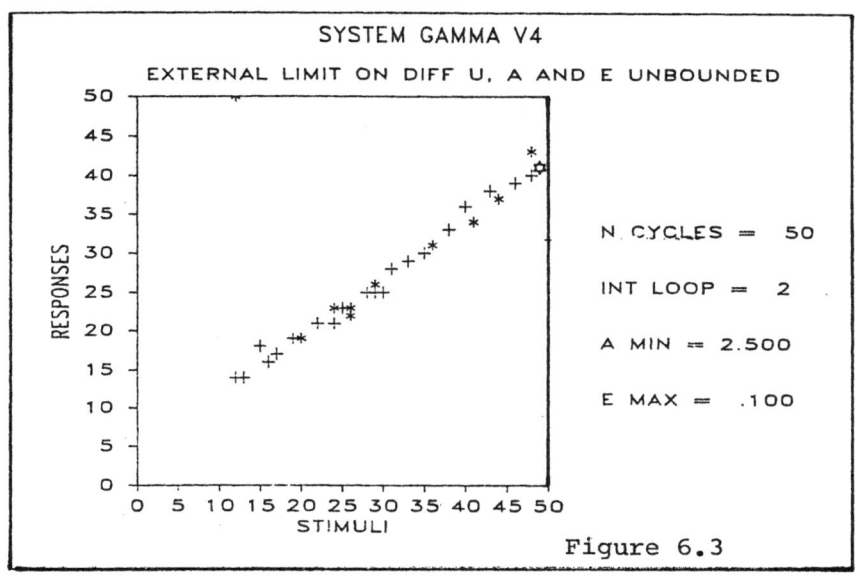

Figure 6.3

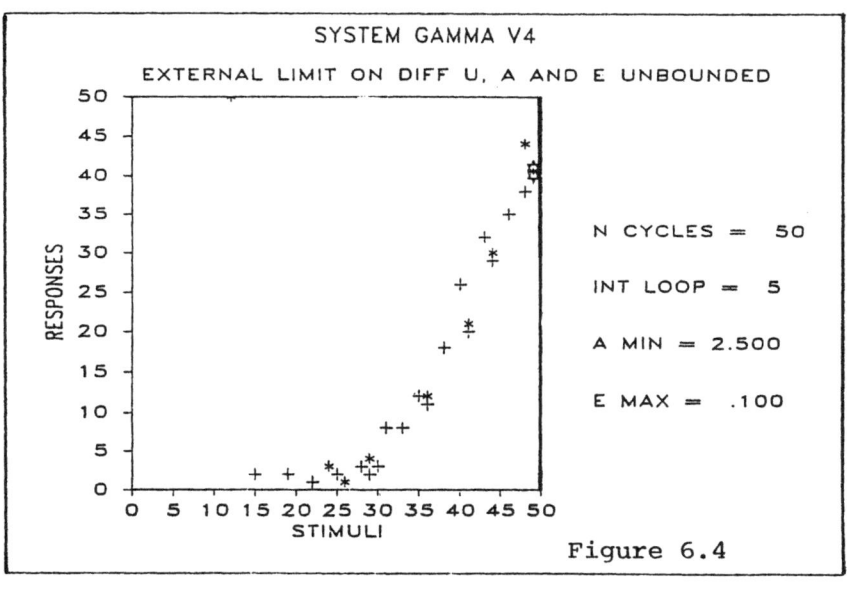

Figure 6.4

drive a car very slowly, and hence to track the induced changes in the highway environment comparably slowly, thus minimising the chance of being surprised or having to make very rapid corrective movements) which can be more restrictive than the organism needs in order to track the output accurately as it varies from trial to trial; there is no effective subsequent internal parameter constraint on either a or on $\Delta^1 a$, which would simulate a sort of inability to cope with change.

For very small η the input-output plots are a straight line with a resemblance to a linear regression with a superimposed small gaussian noise distribution, as is commonly assumed in psychological data analysis, when fitting the psychophysical "law" [3.2]. An increasing appearance of noise is created in this plot as $ae \to 1.7$ (See Table 2.1).

If W_a is very low the output is strongly autoregressive; if $W_a = 1$ then the output explodes, consequently the $U \to Y_{obs}$ plot becomes completely uninterpretable. This simply serves to reconfirm that bounds on the input are needed for the process to be stable. This latter case was coded V 5 in an earlier treatment (Gregson, 1984).

Case V 5

This case is introduced rather as an anomaly, to anticipate some-one asking a "But what happens if...?" question. There is some evidence (Chocolle, 1945, Link and Heath, 1975) that observed reaction times are shorter to more intense stimulation, given that a set of stimuli of various intensities is being presented in the situation involved. If a and e are fixed and U is mapped onto the inverse of η, so that the number of recursive loops is fewer for a greater input, and vice versa then results like Figures 6.7 and 6.8 can be obtained.

We make no attempt to reconcile these findings with experimental results precisely because the observed reaction time is not strictly to be equated with processing time inside a single recursive loop embedded within a total psychophysical system. The Γ model is not a model of discrimination judgment, and so no internal reference stimuli, as opposed to a resting level $Y_0(\text{Re,Im}) = (\zeta, \epsilon)$ of the loop, is assumed.

Case V 6

Here e_J is fixed, and thus uncoupled from $\Delta^1 U_J$. There are no bounds on $\Delta^1 a_J$, and

$$\textbf{V6Ax1}: \quad ie_J = e_{max} = const.$$

The critical variable here is ϵ_c; for low values the output is a straight-line regression, the variance taken up by a linear trend model being up to 92% for $a_{min} = 3.2$, $e_c = 0.1$, $\eta = 2$. But with higher e_c the curve turns over and asymptotes to a value below the Y_{obs} maximum (See Figures 6.9

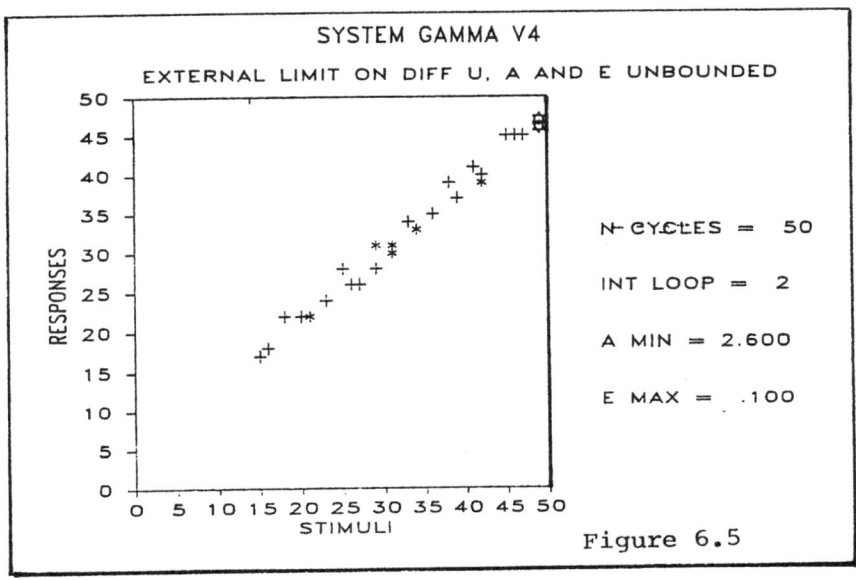

Figure 6.5

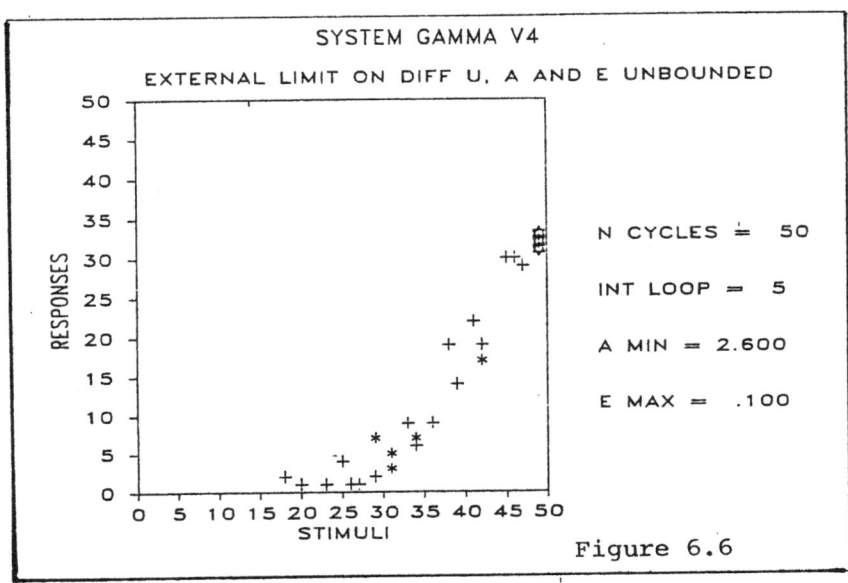

Figure 6.6

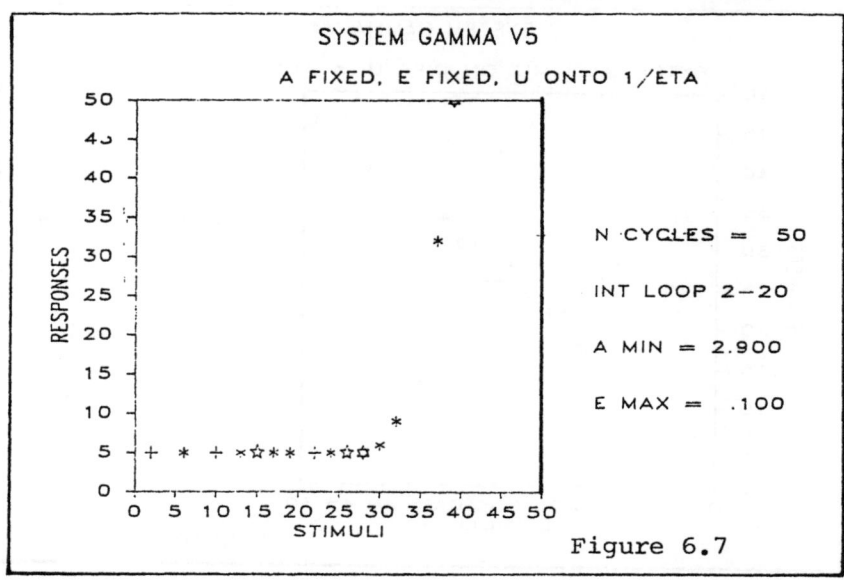

Figure 6.7

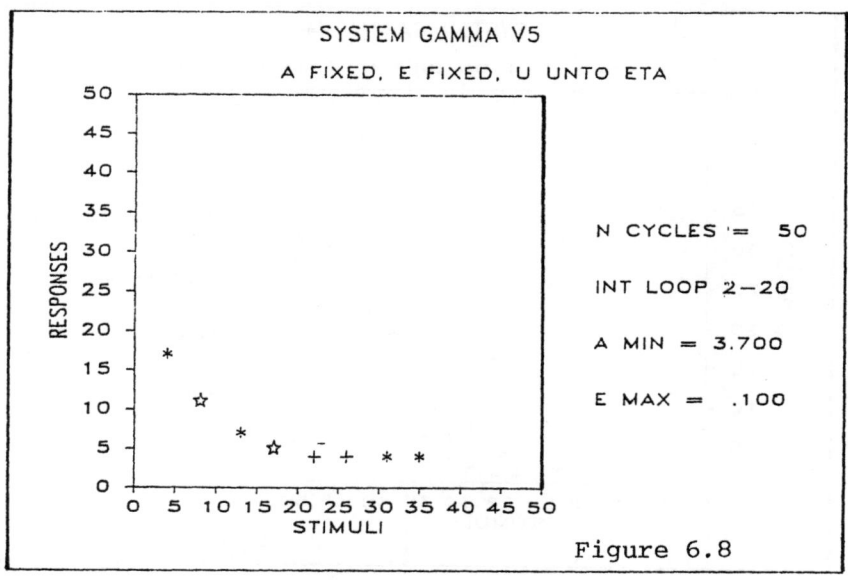

Figure 6.8

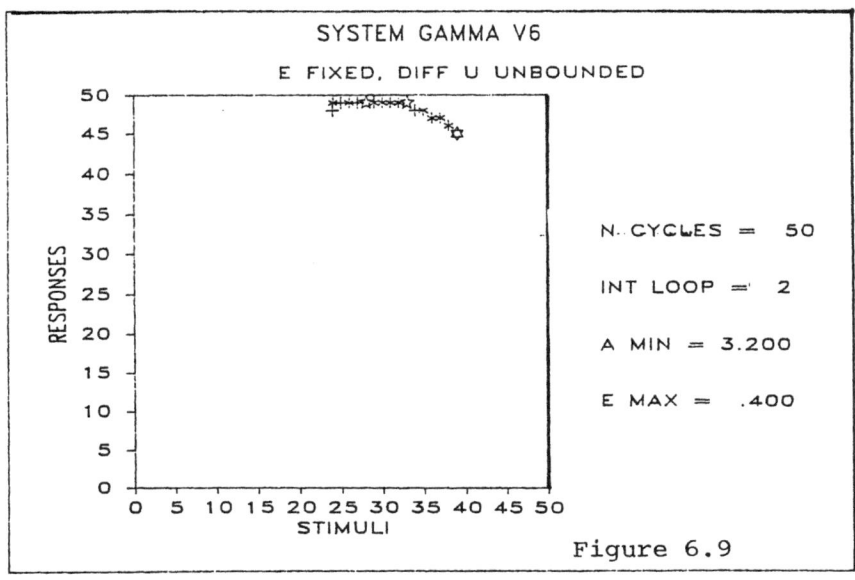

Figure 6.9

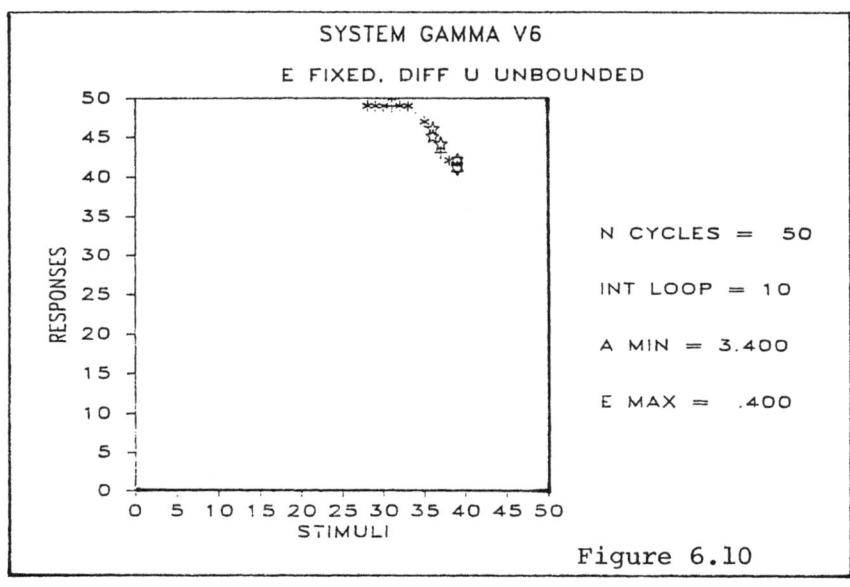

Figure 6.10

and 6.10 for examples). This resembles some results in sensory stimulation in taste and in audition.

Case V 7

This case has been used as a starting point extensively in other chapters, because its output resembles a number of psychophysical experiments, and because it illustrates nicely the wide diversity of input-output relationships generated by small changes in one system parameter, in this case e.

In this variant there are bounds on the rate of change of a, but e is uncoupled from the input and held constant.

V7Ax1 : $U_J \sim rect(z - z_0)$

V7Ax2 : $a_{min} < a < 3.99$

V7Ax3 : $\Delta^1 a \leq \delta_a,$ $\delta_a = const.$

V7Ax4 : $U_j \rightarrow a_J$

V7Ax5 : $e = e_c = const.$

Ax3 is not necessary, but is introduced to reflect the fact that real-world stimulus series are often autoregressive.

The series of cases generated by

$$a_{min} = 3.2, \quad 0.05 \leq e_c \leq .350, \quad \eta = 10$$

is of particular interest, as the role of e in controlling the shape of the psychometric function, sometimes like a CNO, is displayed clearly. Figures 6.11 to 6.18 are samples from the series.

The ogival form is the part of the Γ limit graph in Figure 2.5 before the system begins bifurcations. The constraint in Ax3 keeps V 7 out of the chaotic region; if the system is permitted to enter it without explosions resulting then the value of e_c in Ax4 must decrease a a increases.

Case V 8

Suppose that e is in fact linked to a, so that e is inversely proportional to a. From the known behaviour of the Lyapunov coefficient, fixing e at about 0.5 for $a = 3$, and around $e = .2$ for $a = 6$ should give a system which is stable but exhibits cycles and chaos. These values are not precisely critical. any plot of ae which stays inside the stable bound $\{a, e\}_B$ of Figure 2.7 is admissible, but not necessarily interesting. To examine this case, as much for the curious mathematics as for potential psychophysical realism, a series of 12 plots, in Figures 6.19 to 6.30, are generated with a ranges of 1.0, and a_{min} values of 2.5, (0.5), 6.0. Table 6.1 and Figures 6.19 to 6.30 are to be read together as furnishing a representative picture of Γ V

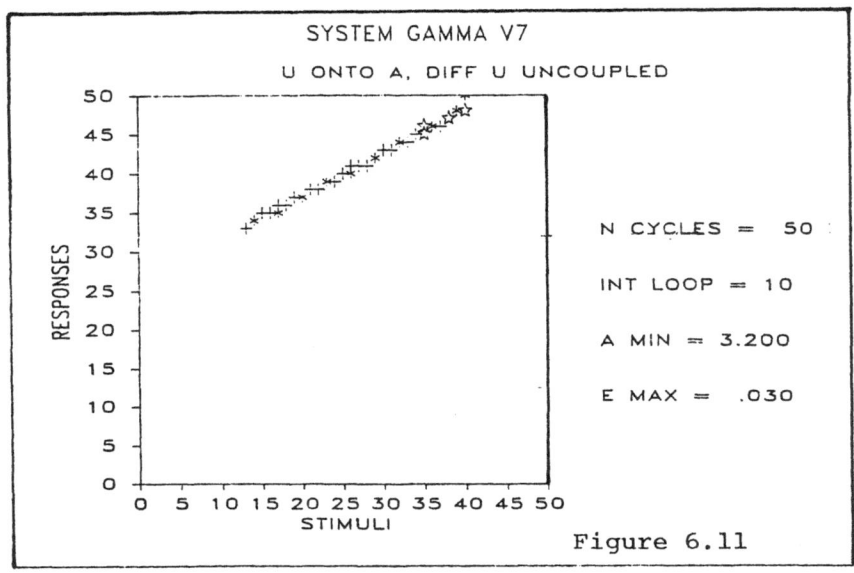

Figure 6.11

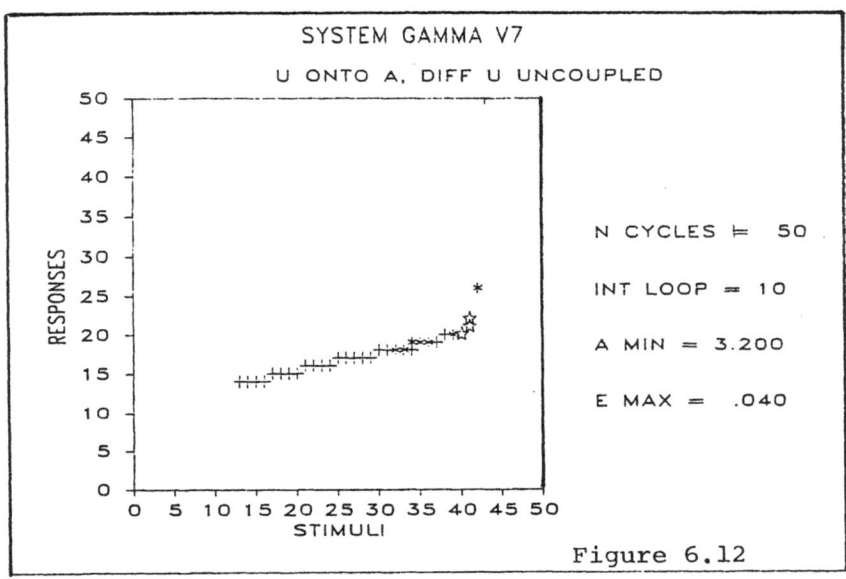

Figure 6.12

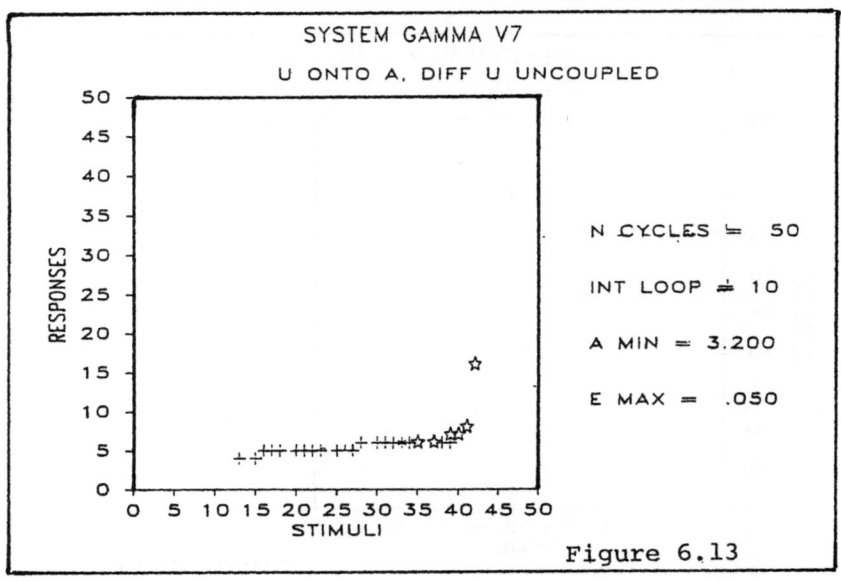

Figure 6.13

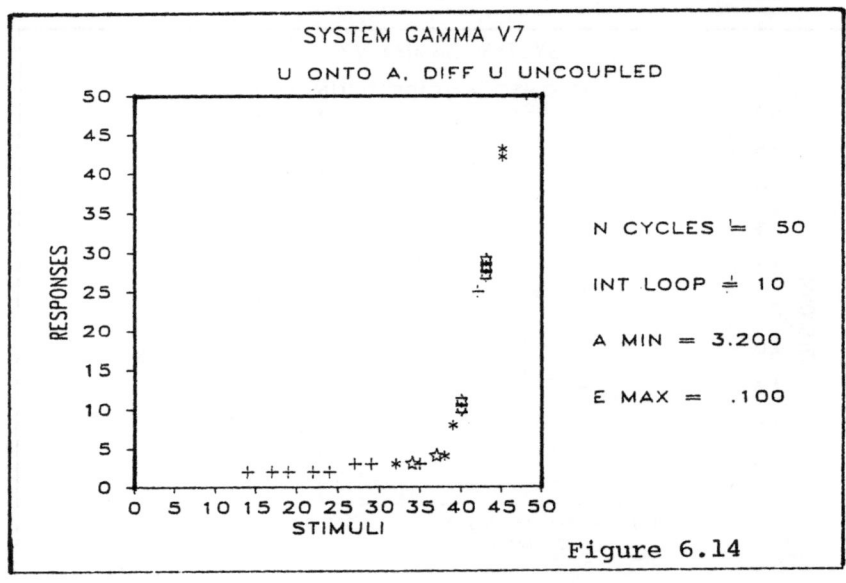

Figure 6.14

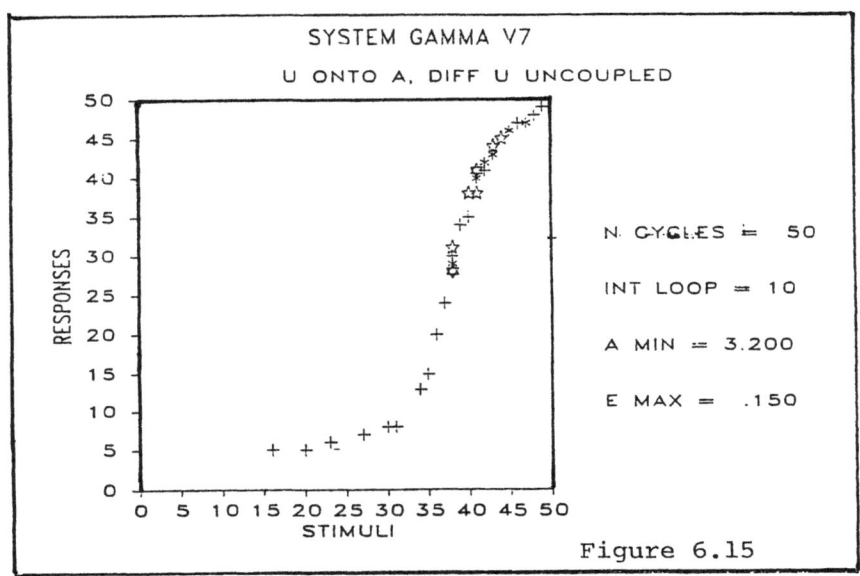

Figure 6.15

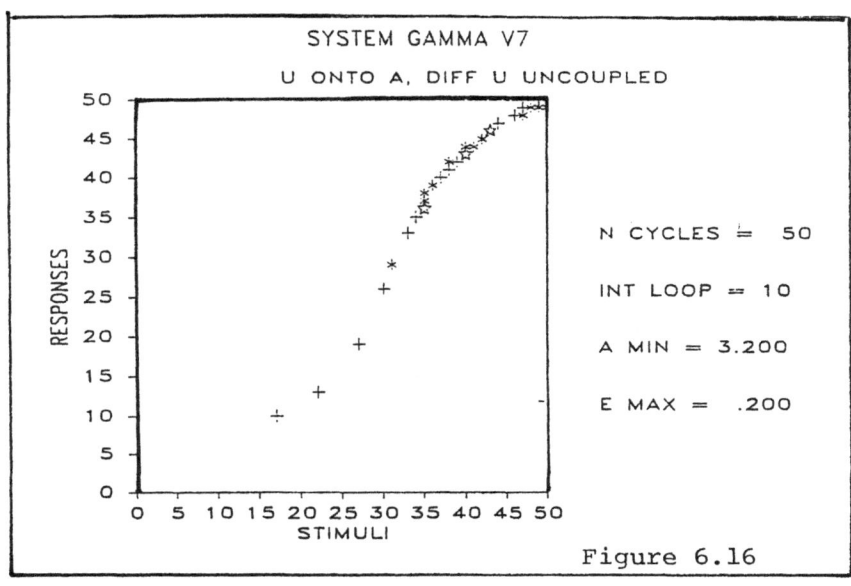

Figure 6.16

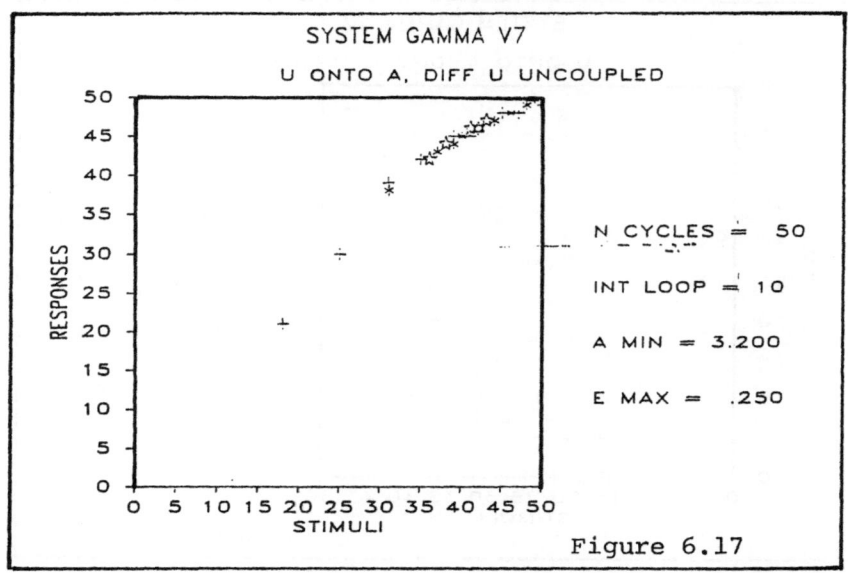

Figure 6.17

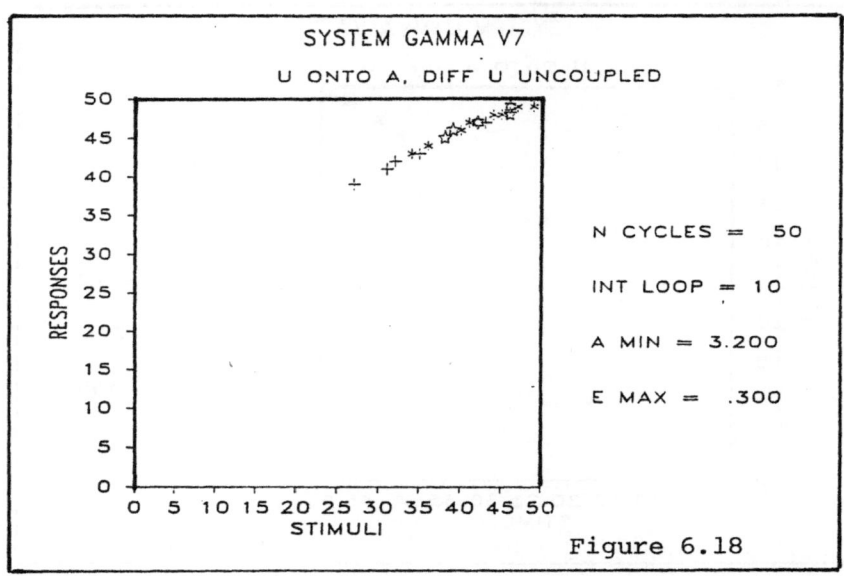

Figure 6.18

8 across the a range of 3.5 to 6.5; only the lower ranges have any potential psychophysical interpretation under usual circumstances.

In Table 6.1 the values H(Re) and H(Im) are based on the information measure

$$\mathbf{H} = -\sum_{i=1}^{20} p_i \cdot log_2(p_i) \qquad [6.1]$$

by dividing the output distribution into 20 segments of equal width, separately for Re and Im components. The U input is rectangularly distributed and hence H = 4.32 bits ($log_2(20)$ = 4.32). The entropy degradation is thus seen to be usually greater, for the Im part H(Re) rises and then falls with increasing a_{min}, whereas H(Im) is erratic. The CNO% value is obtained by calculating (using a NAGF subroutine) the values of a nearest CNO for Y_{obs} based on the observed $mean(Y_{obs}(\text{Re}))$ and its associated range, and then linearly regressing the observed form on that nearest CNO form, effectively by putting

$$\hat{Y}_{obs}(U) = \alpha + \beta \cdot CNO(U) \qquad [6.2]$$

and minimising the residual squared error $(Y - \hat{Y})^2_{obs}$.

This gives an additional check on the ready visual inspection provided by the graph of Figure 6.23, namely that the shape is more like an ogive than any other of the Real plots. The point of this CNO fitting exercise is to underline that a form which is commonly taken to approximate to a CNO can equally well be taken to approximate to predictions derived from Γ V 8, or Γ V 7 at a lower input range.

It can be seen from Table 6.1 and Figure 6.23 that an approximation to a CNO associated with high intensity stimulation is predicted from Γ V 8, but only over a narrow range of inputs before a most confused and confusing pattern emerges.

As the graphs are, for presentation, all scaled so that they fill the 50 × 50 square, the scale in U and Y units varies from plot to plot. The consequences of this are particularly noticeable in the Imag series, which displays the strong influence of ae on the dynamics of the Imags components in the ranges where the system is chaotic.

The range of Y(Im) in Figure 6.20 is about 1×10^{-8}; there is evidence of cycling. What is scaled as coordinates 12 - 36 in Response units in Figure 6.20 shrinks to coordinates 39 - 41 in Figure 6.22; the left half of Figure 6.22 corresponds to the right half of Figure 6.19, and analogously as the a_{min} values are progressively increased in Figures 6.24, 6.26, 6.28, and 6.30. The system goes into quite wild fluctuations about $a = 4.5$ and the input-output function bears no resemblance to a CNO; it is governed by the complicated dynamics shown in Chapter 2. The only possible psychological significance of this section might be to model some sorts of clinical pathology.

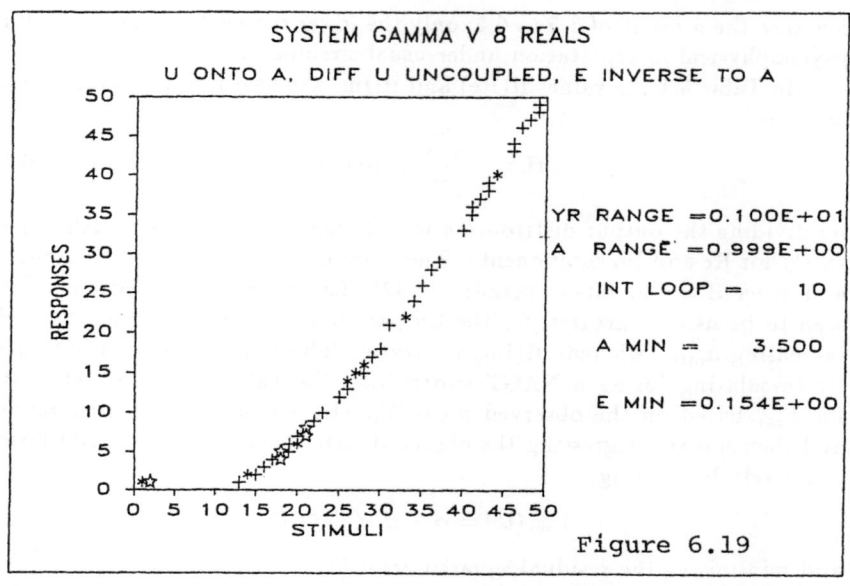

Figure 6.19

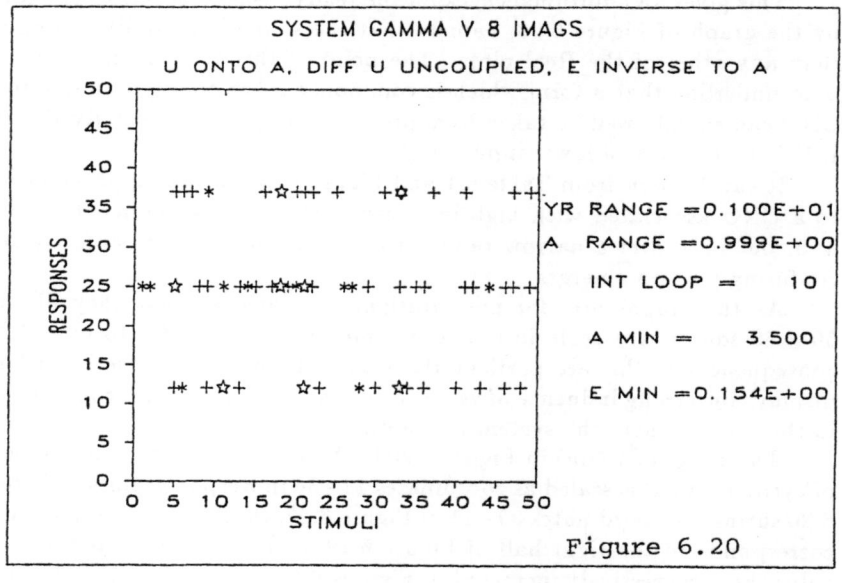

Figure 6.20

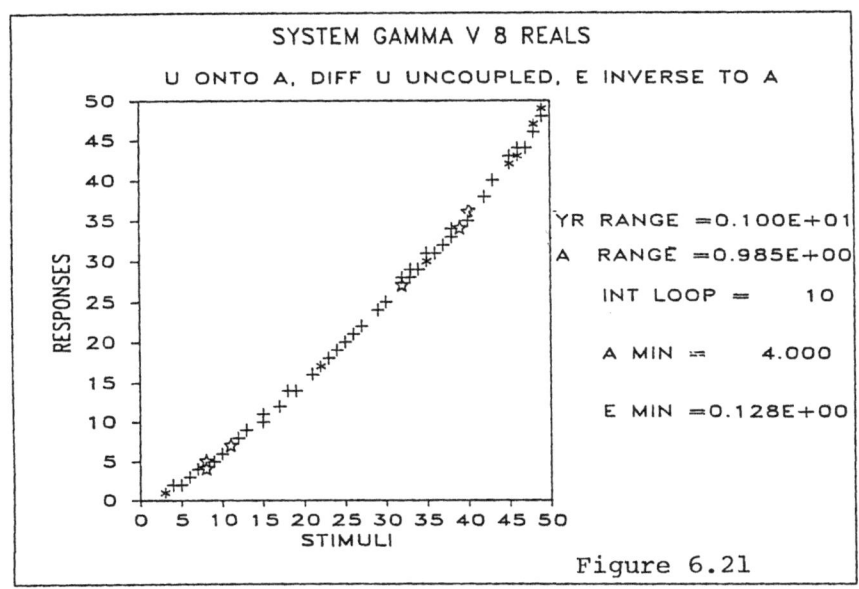

Figure 6.21

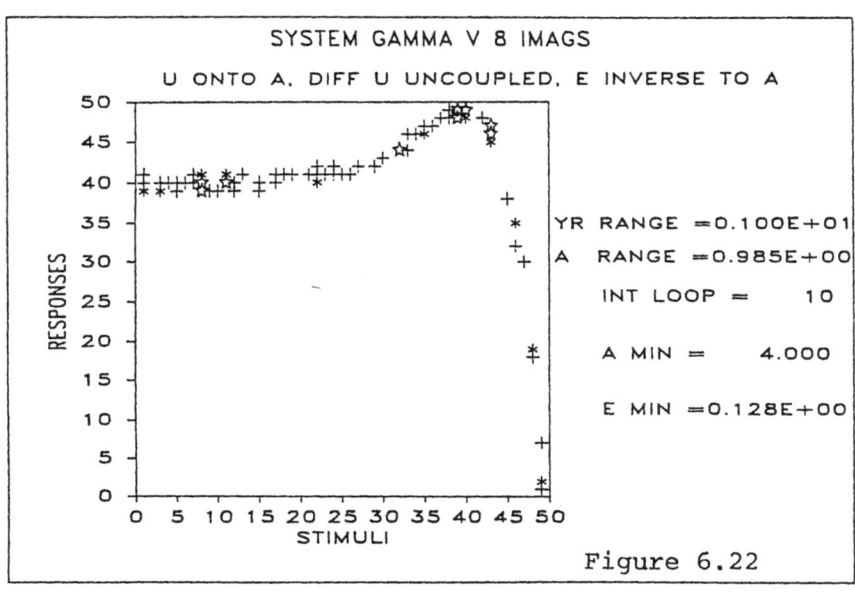

Figure 6.22

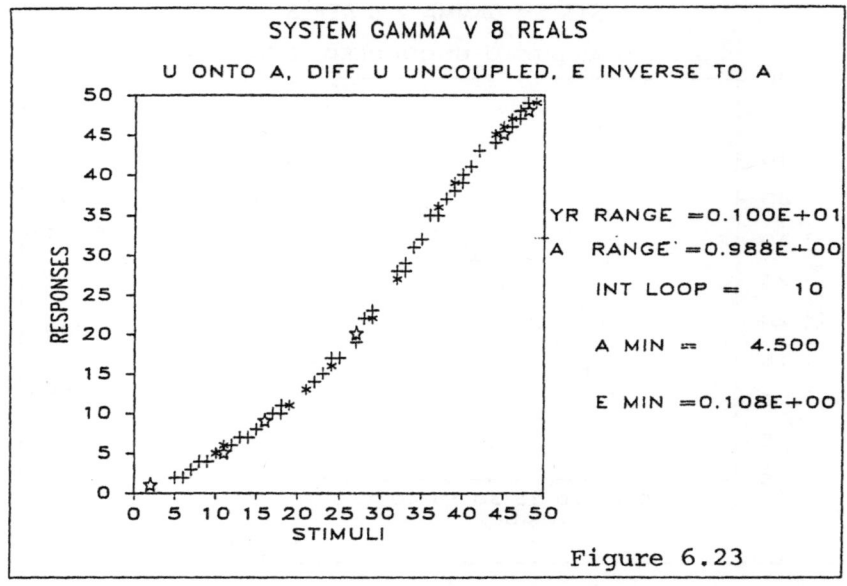

Figure 6.23

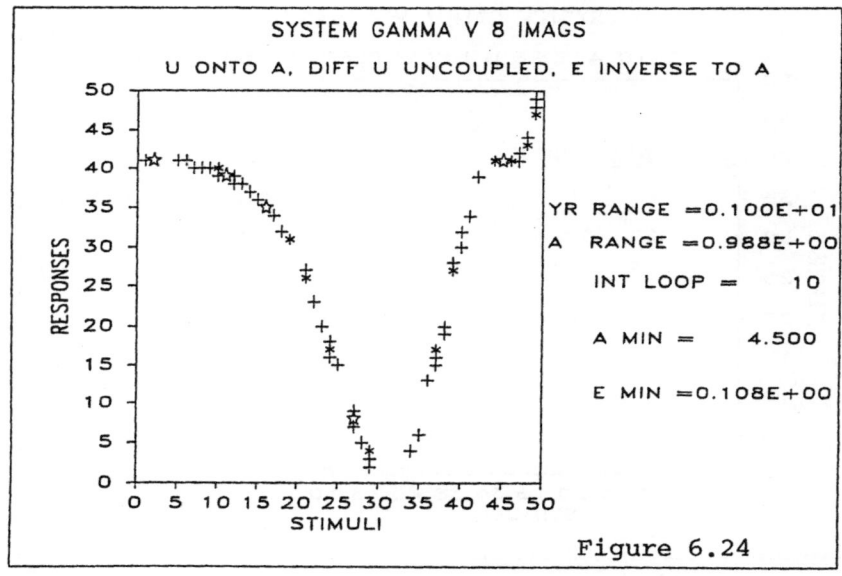

Figure 6.24

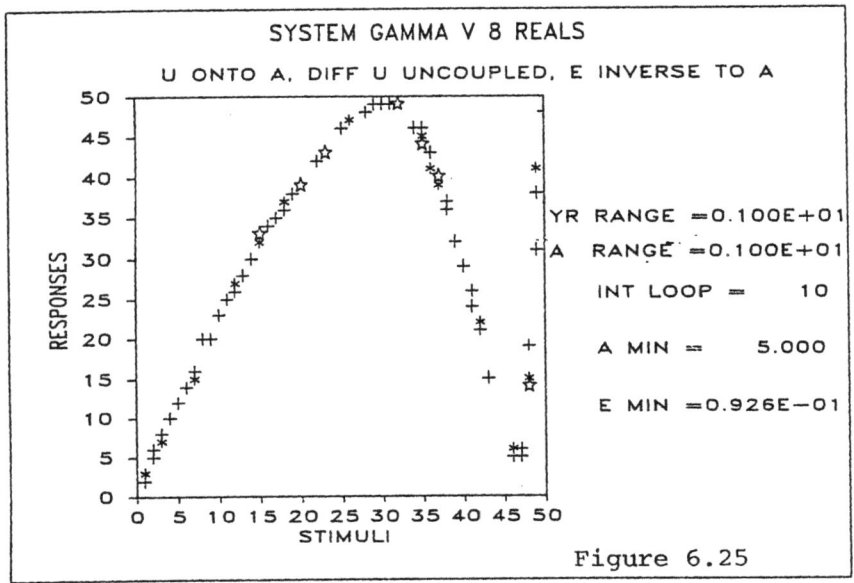

Figure 6.25

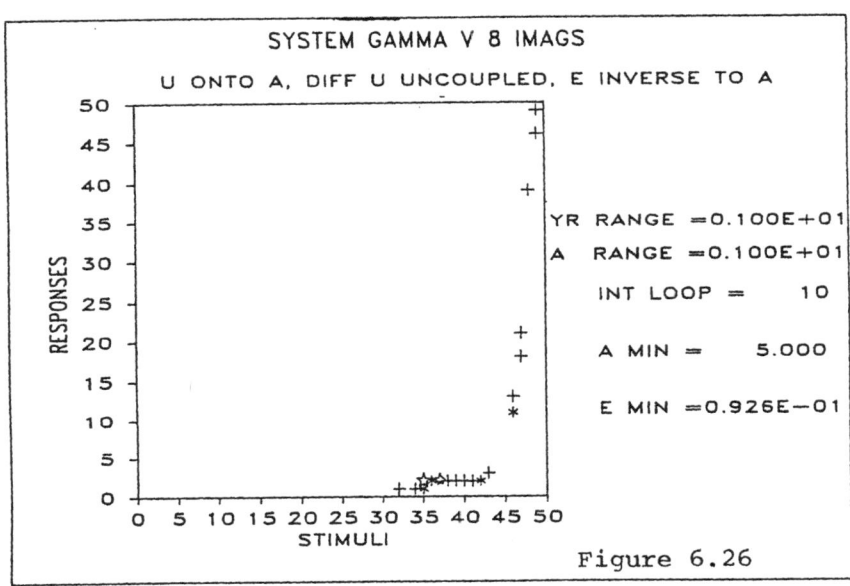

Figure 6.26

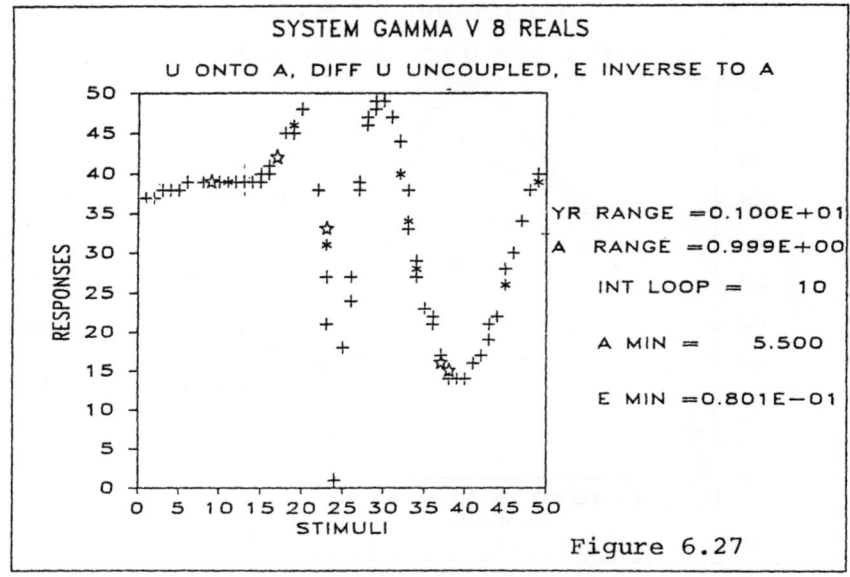

Figure 6.27

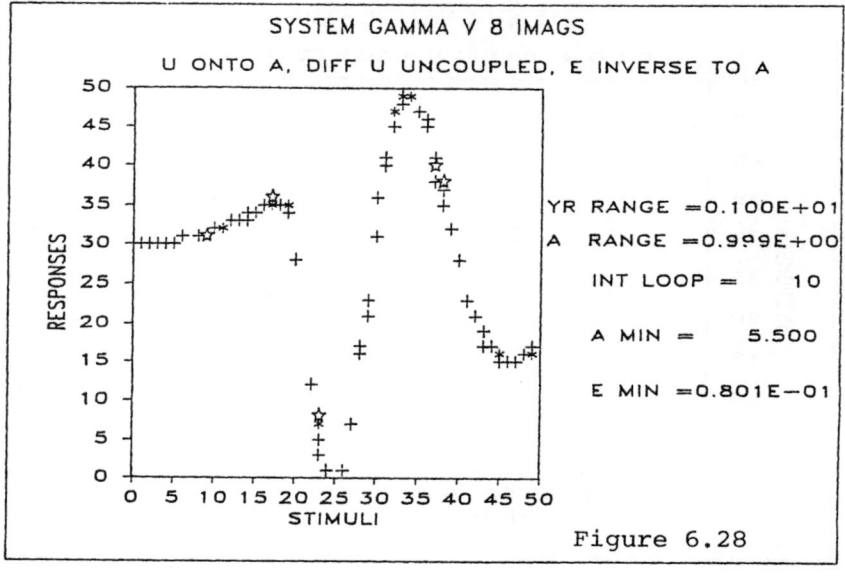

Figure 6.28

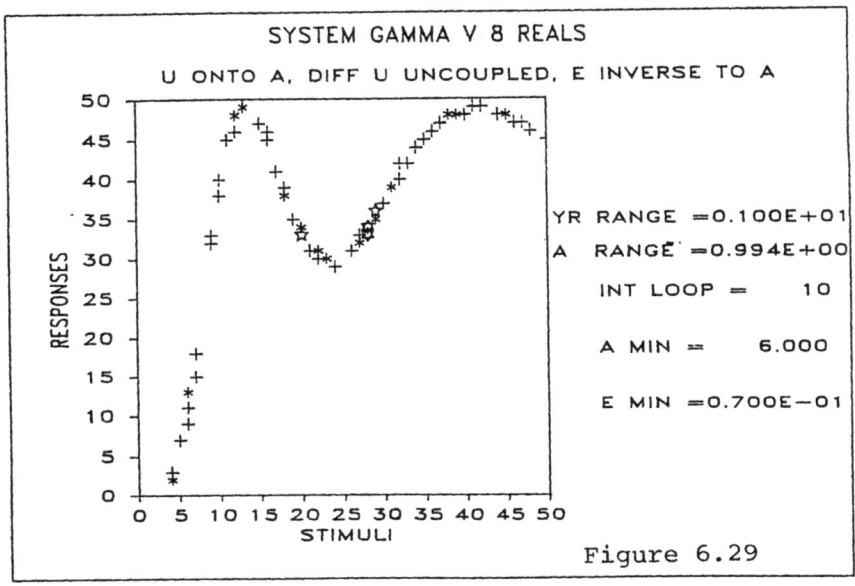

Figure 6.29

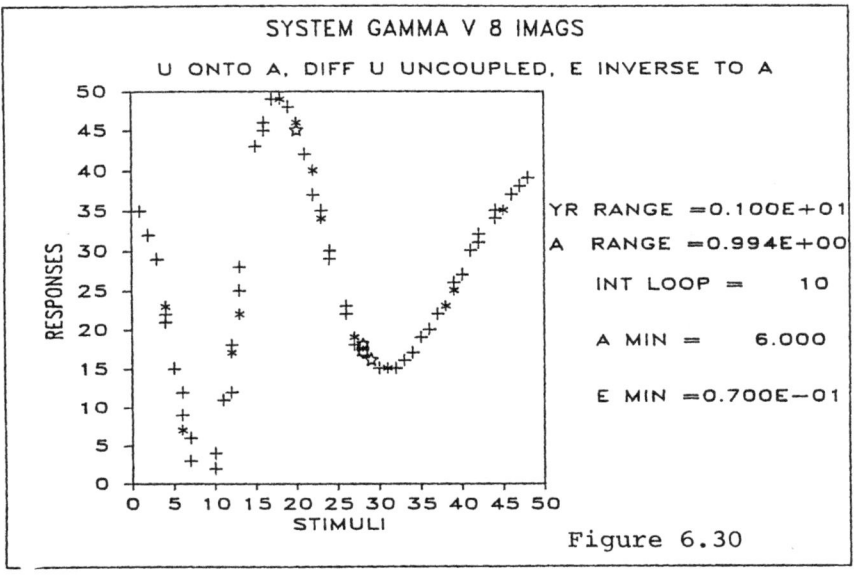

Figure 6.30

Table 6.1

Output Characteristics of Γ V 8 for various a_{min}

a_{min}	min $Y(\text{Re})$	max $Y(\text{Re})$	$H(\text{Re})$	min $Y(\text{Im})$	max $Y(\text{Im})$	$H(\text{Im})$	regress CNO%
3.50	.646	.695	2.712	10^{-8}	$> 10^{-8}$	1.762	24.7
4.00	.674	.743	3.192	$-.134 \times 10^{-6}$	$.316 \times 10^{-7}$	1.954	12.3
4.50	.709	.834	3.068	$-.101 \times 10^{-5}$	$.201 \times 10^{-6}$	2.806	75.4
5.00	.765	.847	3.226	$-.150 \times 10^{-5}$	$.116 \times 10^{-2}$	.739	–
5.50	.672	.906	2.767	$-.174 \times 10^{-2}$	$.111 \times 10^{-2}$	2.816	0.9
6.00	.492	.906	2.509	$-.174 \times 10^{-2}$	$.111 \times 10^{-2}$	3.014	–

Case V 9

This case returns attention to the question of bimodality. The constraints are:

V9Ax1 : $U_{min} < U_J < 3.94$

V9Ax2 : $U_J \to a_J$

V9Ax3 : $a_J = a_{min} + rct(z)$

V9Ax4 : $ie_J = \Delta^1 U_J \cdot W_e$

where W_e is a scalar multiplier.

This case exhibits a slight bimodality; a response level at the lower range of stimulation which is greater than that associated with slightly more input (see Stimulus coordinates 20 - 25 in Figures 6.31 to 6.33) In some ways this is an exaggerated version of Γ V 2; these lower limits of a u-shaped stimulus-response relationship are of potential interest because the pattern depicts a system where resting state (no stimuli) activity levels are higher than those aroused with low inputs. It resembles the phenomenon in Parkinson's Disease where the patient has a tremor in his/her hands which is only suppressed when there is a task to be performed which demands limb movement and dexterous action. The responses which are evoked when the tremor is suppressed are often clumsy, that is, almost all-or-none responding.

Other cases can obviously be readily devised, for example by lagging the coupling of $\Delta^1 a = \Delta^1 U_{j-1} \cdot W_a'$ relative to $a = W_a \cdot U$, where W_a and W_a' are weights. This sort of construction can produce *one $\to$ many* $U \to Y_{obs}$ relationships which it would be difficult to discern in experiments. The few cases given in this chapter are not all that have been explored by simulation,

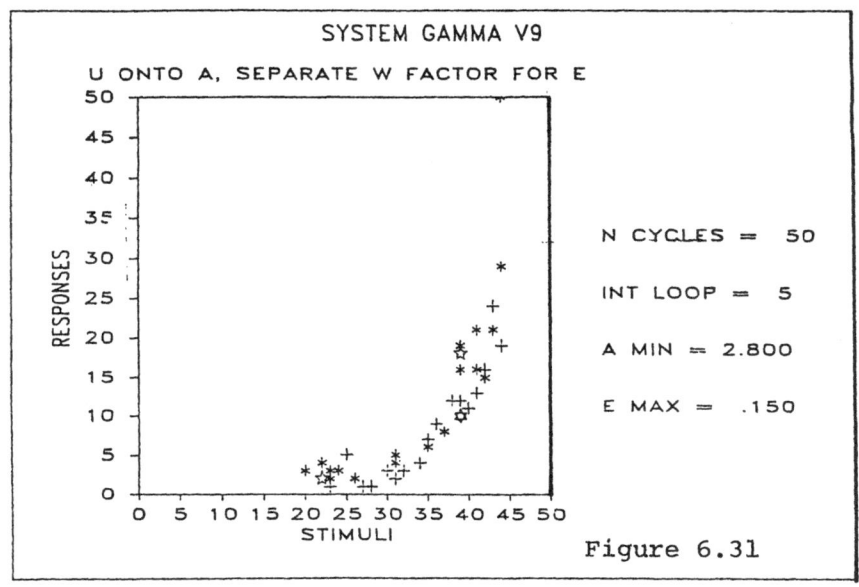

Figure 6.31

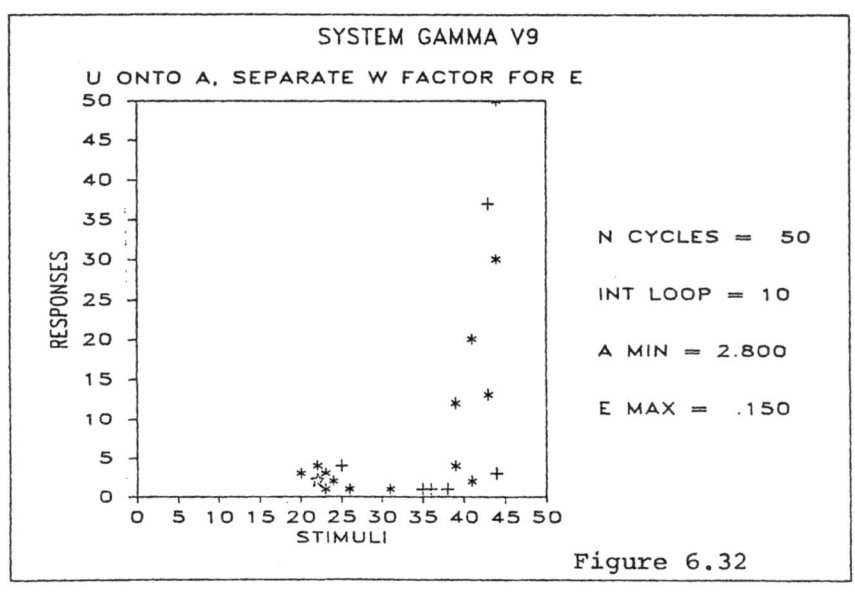

Figure 6.32

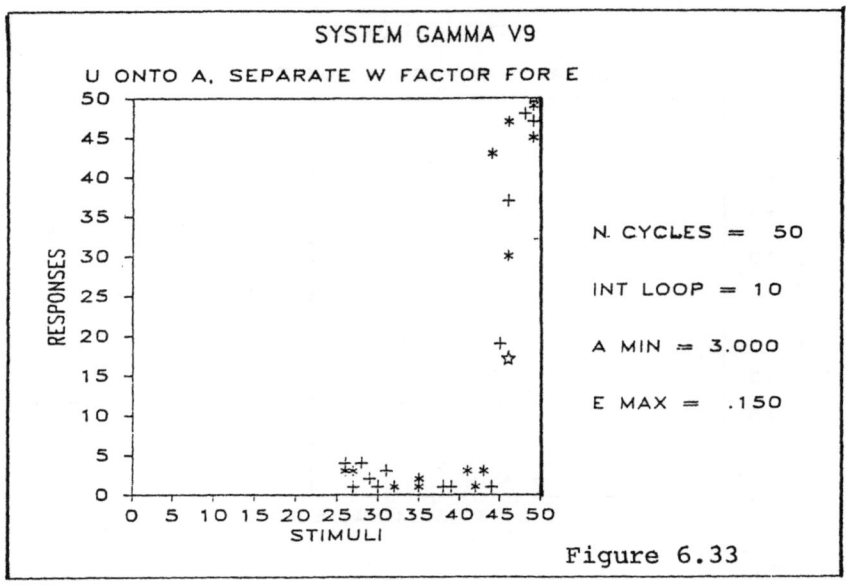

Figure 6.33

they are sufficient to show that forms can readily be constructed to mimic a diversity of real-world phenomena, or to predict their occurrence under strictly limited input and system parameter combinations.

References

Chocolle, R. (1945) Variation des temps de réaction auditifs en fonction de l'intensité à diverses fréquences. *L'Année Psychologique, 41,* 65 - 124.

Gregson, R. A. M. (1984) Behaviour of a system with gain and pure delay filters incorporating a nonlinear difference feedforward loop as a generalized psychophysical model. Unpublished seminars: Federal Republic of Germany, German Democratic Republic, and Sweden.

Gregson, R. A. M. (1986) Qualitative and Aqualitative Intensity Components of Odor Mixtures. *Chemical Senses, 11,*

Helson, H. (1947) Adaptation level as a frame of reference for prediction of psychophysical data. *American Journal of Psychology, 60,* 1 - 29.

Krantz, D. H. (1972) A theory of magnitude estimation and cross-modality matching. *Journal of Mathematical Psychology, 9,* 168 - 199.

Link, S. W., and Heath, R. A. (1975) A Sequential Theory of Psychological Discrimination. *Psychometrika, 40,* 77 - 105.

7   Cascading of the Loop

What happens when we put a series of Γ loops in series can be interesting and counter-intuitive. We are led to consider this question for two reasons, at least, partly because we know that neural networks can be in cascade, and have long and short loops sharing common pathways for part of the loops (Marczynski, 1986), and partly because it is possible that the reduction of input series of almost-continuously variable signals to output series with long periods of quiescence interrupted by sporadic outbursts, may be produced by cascading loops each of which by itself could not generate the phenomena of interest.

That series of responses exhibit sudden rare but large jumps seems to have been noticed, and has been reported anecdotally to this writer, but little or nothing formal seems to have been published. Rare and unpredictable phenomena are intractable objects of study, but are of obvious interest to the clinical psychologist concerned with abnormalities of sensory function. For the experimenter trying to use the methods of classical psychophysics to determine thresholds, abrupt jumps in a threshold followed by extended periods of stability are a considerable nuisance (Roufs, 1973) because such jumps create both theoretical problems concerning how we should define a threshold, and practical problems concerning how to measure it. Such jumps would also be of comparable nuisance value in a signal detection paradigm, the mean of the noise distribution on the decision axis would exhibit similar saltations.

As we have already observed, peculiar jumps, breaks, and strange response variance distributions happen in the world (Rodenburg, Maas and Stassen, 1981) and are to be expected in nonlinear psychophysics; they are intrinsic to some regions of the system parameter space. It might there-

fore be possible easily to find sufficient conditions for the generation of such
things as threshold jumps, if the phenomena are associated with the complications of nonlinearity, without adding any more parameters or boundary
conditions to the system. We illustrate here how some jumps might readily occur; the calculations are straightforward but laborious if we seek to
investigate all possible modes of generation.

It is desirable to review various cases, in an effort not only to identify
sufficient conditions for the occurrence of jumps, but also by elimination
to see what is the generality of parameter configurations which are associated with jumps. In a nonlinear system the output dynamics can remain
unchanged over a wide hypervolume of the $\{a, e, \eta\}$ parameter space (the
Andronov stability of system theory), so possibly the identification of a
bounded region of the parameter space for which jumps will almost always
occur eventually, is very improbable *a priori*.

Using Γ V 7 gives an autoregressive output series Y_{obs} for a random
input series U, and it is expedient to examine separately the Re and Im
component output series. Time series analyses of Y_{obs} are often intractable
if reliance is placed solely on linear (Box-Jenkins) modelling; some samples we have examined appeared to be nonstationary and nonlinear, and
cannot be fitted with satisfactorily low residuals to low-order ARIMA p, d, q
models. Attempts to fit SETAR (Self-Excitatory Threshold Autoregressive modelling, Tong (1983)) with three states to the Y_{obs}(Re) data were
not satisfactory.

The output of Γ is Y_{obs}(Re,Im), for a U(Re) input. The results of
Chapter 5 suggest that it is plausible to think of Y_{obs}(Re) as a representation of observable output and Y_{obs}(Im) as a representation of internal
activity levels within the system.

In this search for saltation we have cascaded Y_{obs}(Re) and Y_{obs}(Im)
separately back into the difference equation of Γ by mapping the Re or Im
output into a for the next cycle. The two cascades (0 cascades = 1 pass
from U to Y_{obs}, 1 cascade = 2 passes in series, and so on) are generated in
the following diagrams as shown. In the case of cascaded Reals the mapping
is from Reals onto Reals repeatedly; but for the Imaginaries the mapping
is Re to Im to Re to Im, etc. Using $\emptyset$ to indicate discarded components,
we have adopted here the layout given in Flow Diagram 7.1.

The Imag cascading implies that at each step the noise is split off and
refiltered back through the nonlinear recursive loop.

We have not so far considered the case where only a_1 is Re, and
$a_2, ..., a_n$ are made complex, which is a possibility (Mandelbrot, 1980
does this with the logistic equation [2.1]). It may not in fact be necessary
to explore this, as $O(\text{Im}) = 10^{-8} \cdot O(\text{Re})$ for Y_{obs}(Re,Im), so that feeding Y_n
fully normalised on both Re and Im components back into a_{n+1} would be

Flow Diagram 7.1

Real cascade :

$$a_{1Re} \rightarrow \begin{cases} Y(\text{Re}) \rightarrow a_{2Re} \rightarrow \begin{cases} Y(\text{Re})_2 \rightarrow a_{3Re} \rightarrow \begin{cases} Y(\text{Re})_3 \rightarrow a_{4Re} \\ Y(\text{Im})_3 \rightarrow \varnothing \end{cases} \\ Y(\text{Im})_2 \rightarrow \varnothing \end{cases} \\ Y(\text{Im}) \rightarrow \varnothing \end{cases}$$

Imag cascade:

$$a_{1Re} \rightarrow \begin{cases} Y(\text{Re}) \rightarrow \varnothing \\ Y(Im) \rightarrow a_{2Re} \rightarrow \begin{cases} Y(Re)_2 \rightarrow \varnothing \\ Y(Im)_2 \rightarrow a_{3Re} \rightarrow \begin{cases} Y(re)_3 \rightarrow \varnothing \\ Y(Im)_3 \rightarrow a_{4Re} \end{cases} \end{cases} \end{cases}$$

almost like feeding back $Y_n(\text{Re})$, whereas feeding back $Y_n(\text{Im})$ separately normalized as is done here means that it is not entirely swamped by $Y_n(\text{Re})$.

A fully normalised $Y_n(\text{Re,Im})$ could yield a series like the cascaded reals, if the noise generated on the first pass through the feedforward loop is split off and lost, so that only the Re component maps onto a on the second pass.

Simulations

Cascades 0,1,2 and 6 for Re and Im in Γ V 7 with $a_{min} = 2.67$, $e = .05$, $\eta = 5$, with N = 300 replications of the loop were run. Table 7.1 gives the coefficients of pure AR(7) models fitted to each and some comments on the residuals; program BMDP2T was used for the time series analysis.

A brief qualitative description of sequences of 300 trials for Re and Im for Γ V 7, where C = the number of cascades, follows:

To see if the runs of $Y_{obs}(\text{Im})$ occur in other slightly different versions of Γ, simulations of V 7 have been run with $a = 3.0$, $e = .50$, $\eta = 5$, for C0 and C6, which is far removed from the simulations in Table 7.1. For C0 we get that $Y_{obs}(\text{Im})$ stays mainly at one level, but jumps to two adjacent levels in a quasiperiodic fashion for short bursts: the associated $Y_{obs}(\text{Re})$ are autoregressive but variable. For C6 the $Y_{obs}(\text{Im})$ degenerate into a single fixed value with rare outliers, the corresponding $Y_{obs}(\text{Re})$ becomes a constant high value. This case does not appear to resemble any real-life data that have so far come to our attention, so it is held over.

Table 7.1

AR(k), k = 1,....,7 coefficients for Γ V 7

	Err#	k 1	2	3	4	5	6	7
Reals								
C0	+1	.989*	.006	-.106	-.021	-.086	.029	-.108
C1	+1	.984*	.339*	-.119	-.176*	-.022	-.128*	.103
C2	+1	.983*	.353*	-.075	-.028*	-.107	-.051	.005
C6	++1	.992*	.489*	-.265*	-.018	-.216*	-.067	-.276*
Imags								
C0	++2	.129*	.939*	.148*	-.191*	.121	.002	.013
C1		.041	.711*	.227*	.114	.201*	.091	.076
C2	++2	.398*	.851*	.315*	.163*	.036	-.121	.020
C6	++2	-.057	.890*	.095	-.021	.265*	.025	.195*

In Err, a case where residuals are outside tolerance limits for one or more lags, as given, is shown by a +.
* $t \geq 1.96$.

ReC0: Autoregressive with fluctuations resembling the original random walk input.

ReC1: Minor and major outbursts on an irregular baseline.

ReC2: See Figure 7.1.   ReC6: See Figure 7.2.

ImC0: Short runs and oscillations through 2 or 3 levels; resembles chaos and intermittency.

ImC1: Brief sustained single value runs, and periods of oscillation usually between two values, not always the same two.

ImC2: See Figure 7.3.   ImC6: See Figure 7.4.

Recalling that Γ V 1 is defined by

$$U \rightarrow a, \qquad \Delta^1 a < \mathbf{C}, \quad \mathbf{C} = const.,$$

$$\Delta^1 U \rightarrow (0, e),$$

this case serves as a prototype as it appears to be a realistic model of some human sensory performance. Simulating a set of results with $a_{min} = 2.6$, $\mathbf{C} = .05$, $\eta = 5$, for 300 trials, which corresponds exactly to Table 7.1, the autoregressive coefficients of Table 7.2 were obtained.

Patterns for a case of interest are shown in Figures 7.5 and 7.6. Qualitatively the picture resembles that from Γ V 7 , but is much less clear; the $Y_{obs}(\mathrm{Re})$ series retains some variability at its base level, and $Y_{obs}(\mathrm{Im})$ shows weak short runs at a single value, but does not show the long runs of V 7, and it dances between values.

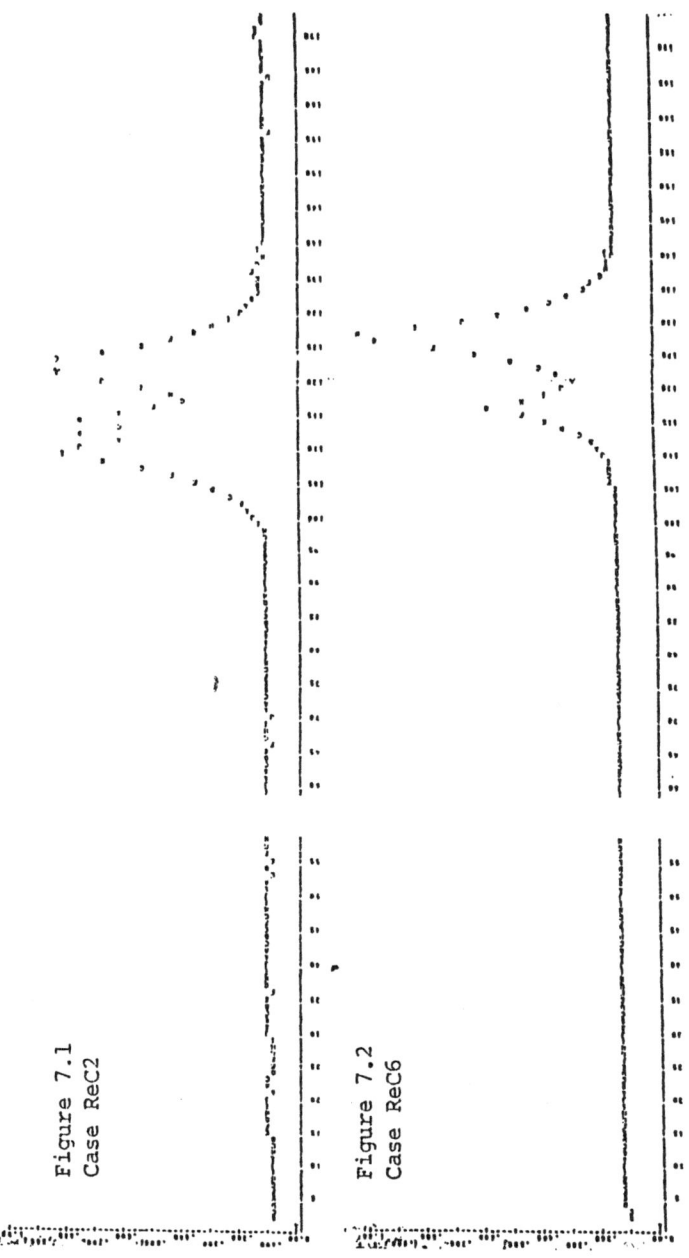

Figure 7.1
Case ReC2

Figure 7.2
Case ReC6

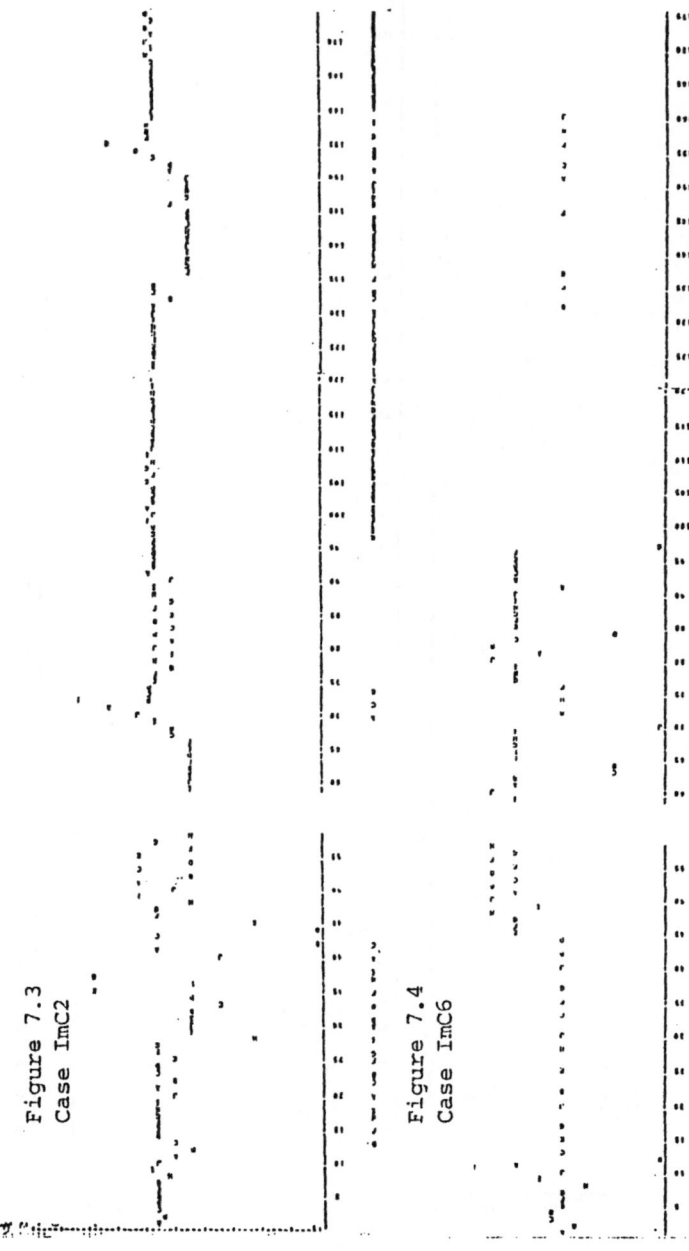

Figure 7.3
Case ImC2

Figure 7.4
Case ImC6

Table 7.2

AR(k), k = 1,....,7 coefficients for Γ V 1

	Err#	1	2	3	4	5	6	7
					k			
Reals								
C0	++1	.981*	-.003	-.043	-.086	-.011	-.007	-.029
C1	+1	.959*	.165*	.083	-.176*	-.050	-.005	.092
C2		.953*	.242*	-.016	-.163*	-.047	-.029	-.032
C6		.617*	.412*	.396*	.107	.189*	-.144*	-.011
Imags								
C0	+4	.342*	.429*	.281*	.262*	.234*	.175*	.197*
C1	+9	.473*	.210*	.138*	.318*	.198*	.121*	.048
C2		.458*	.202*	.144*	.044	.298*	.219*	.237*
C6	+4,5	.37*	.041	.212*	.501*	.362*	.242*	.379*

It may be concluded that the mapping W of input U series character-istics into the nonlinear system parameters is critical for a clear demonstra-tion of the threshold jumping phenomenon; it only appears to be generated in some restricted cases, where the restriction is both in terms of the pa-rameters $\{a, e, \eta\}$ and in terms of the mapping W (V 7 versus V 1).

Obviously we have not explored extensively the whole range of possible mappings that were mentioned in Chapter 6, and so there is no formal proof or demonstration that other regions of the parameter space exist which might generate jumps. The identification of the necessary and sufficient conditions for the saltations is outside the scope of this monograph. Two problems arise which deserve caution and attention:

(i) It has been assumed, from inspection of plots of Y_{obs}(Re,Im) over time, and from results on simulating the subjective Weber function, that it is reasonable under a diversity of conditions to interpret Y_{obs}(Re) as the observable output of the system, corresponding to the quantifiable aspects of the responses made by a human subject in a scaling or judgment task, and that Y_{obs}(Im) corresponds to the internal system's own separate output which is statistically usually interpreted as residual noise, and which can generate threshold effects by acting as a variable background level masking signals, or even by producing some false positive responses from time to time in a detection paradigm. Why the mapping (Re, Im)↦ (Observables, Noise) is so facilely made is not entirely clear.

(ii) The observable responses, and the noise at output, have been as-sumed to arise at the same cascade level; that is, if Y_{obs}(Re) is from Cn, then equally the associated Y_{obs}(Im) is from Cn. It could however be that this assumption is both unnecessary from a modelling perspective, and false empirically in the nearest analogous neurological systems. If Y_{obs}(Re) is

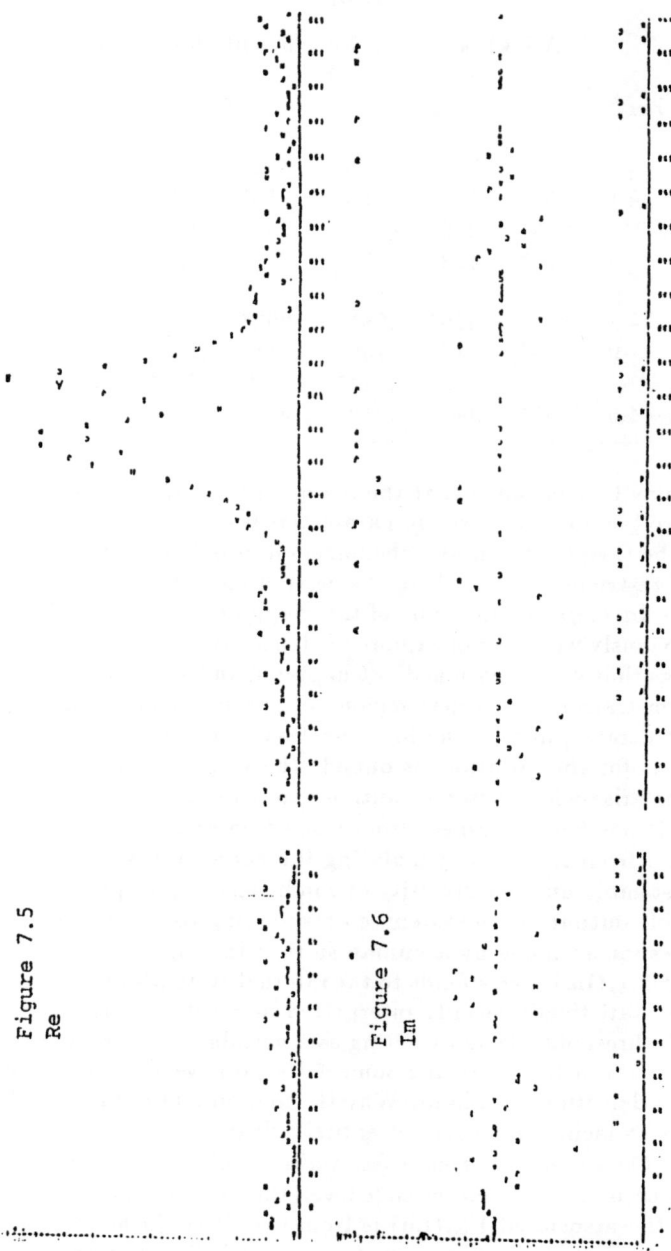

Figure 7.5
Re

Figure 7.6
Im

from C0 or from C1, and Y_{obs} (Im) is from C5 or beyond, then we would get under V 7 an input-output function that looks like a psychophysical CNO for weak stimuli, and at the same time the jumping threshold phenomenon. It is neurophysiologically quite possible to have perseverating reverberating circuits (Petsche *et al*, (1974) is one of the earlier papers reporting such general activity in masses of cortical tissue), and an apparently deterministic output might be preferentially attuned with uncascaded paths. What is mathematically uncascaded is simply a path with linear transformations downstream to the output after the one nonlinear loop of Γ, which again is possible; for example linear transformations from activity patterns in the visual cortex to overt perceptual responses do occur.

It remains to note that we could redefine a threshold as an ordered set of discrete activity levels, between which the system could stochastically jump. Actually in Γ the jumps are generated deterministically and not stochastically, but an external observer would have considerable difficulty in determining this, if the system structure is not known. These jumps are reminiscent metaphorically of an electron jumping between orbits; what we would wish to know here is the expected distribution of interjump intervals in time, which is a time series possibly representable by a semi-Markov chain.

Variations in human sensory and vigilance performance on a 24-hour or 12-hour biological rhythm are, of course, well known, and could account for changes in threshold at different times of the day. The question still has to be put, how do the biological changes in the organism that are rhythmic operate on the parameters of the sensory system? Shifting a_{min} (as a function of arousal) could produce something like threshold changes, as shown in the different results for V 7 reported here.

The idea of diffusion through a neural network which is implicit in the cascading of both the Re and Im components, but more so in the cascaded Im noise process, is of course already well-known in mathematical neurobiology, and has been the subject of many nonlinear models which are basically either diffusion processes or Volterra equations, as reviewed by an der Heiden (1980). Earlier a formal model of transport and control in neuron networks was given by Swigert (1970).

References

An der Heiden, U. (1980) *Analysis of Neural Networks, Lecture Notes in Biomathematics No. 35,* Berlin: Springer-Verlag.

Mandelbrot, B. B. (1980) Fractal Aspects of the Iteration of $z \mapsto \lambda z(1-z)$ for Complex λ and z. *Annals of the New York Academy of Sciences, 357* , 249 - 259.

Marczynski, T. J. (1986) A model of brain function. *In* Rebert, C. S.

(Ed.) *Neural Anatomy, Chemistry and Event-related Brain Potentials: An Approach to Understanding the Substrates of Mind.* Supplementum to *Electroencephalography and Clinical Neurophysiology.*

Petsche, H., Prohaska, O., Rappelsberger, P., Vollmer, R., and Kaiser, A. (1974) Cortical Seizure Patterns in Multidimensional View: the Information Content of Equipotential Maps. *Epilepsia, 15,* 439 - 463.

Rodenburg, M., Maas, A. J. J., and Stassen, H. P. W. (1981) Thresholds for the Perception of Rotation: Variability, Psychometric Curves, and Comparison with Hearing Thresholds. *Biological Cybernetics, 42,* 23 - 28.

Roufs, J. A. J. (1973) *Dynamic Properties of Human Vision.* Proefschrift, Technische Hogeschool Eindhoven, The Netherlands.

Swigert, C. J. (1970) A Mode Control Model of a Neuron's Axon and Dendrites. *Kybernetik, 7,* 31 - 41.

Tong, H. (1983) *Threshold Models in Non-linear Time Series Analysis. Lecture Notes in Statistics No. 21.* Berlin: Springer-Verlag.

8 Elementary Identification of Variants and Parameters

Variation in Information Output

The input of U as a rectangular probability distribution to a Γ variant does not necessarily result in an output Y_{ob}, which is also rectangular; almost always it is not so. Hence the output entropy is less than the input entropy, because the input entropy is maximum by definition. As the behaviour of a system under maximum uncertainty is a limiting case of both theoretical and empirical interest, it is a starting point for simulations of very wide generality. A gaussian distribution of inputs is also, of course, a special case which has been espoused in Signal Detection Theory, but here a rectangular distribution has been most often used for three reasons: (i) it is easy to create with standard programming languages, (ii) as it is maximum entropy it creates the widest range of input conditions equiprobably and thus imposes a maximal load on the system as an adaptive process model, and (iii) it comes close to what is common laboratory practice in psychophysical experiments, and hence any comparison between the maximum entropy case and one of less entropy is indirectly a test of the ecological validity of the laboratory situation, and a prediction of how behaviour will differ in moving from the laboratory to the field. We can only study how a nonlinear system degrades entropy as a function of the system parameters $\{a, e, \eta\}$ if the input is not already in some imperfectly understood form locally degraded.

Under some circumstances $\Gamma V1$ degrades the input entropy appreciably, and it can be shown that all three of a, e, η contribute to this information loss. It is therefore necessary, even if only for illustrative purposes, to explore a plausible range of the parameter space in all three of the variables involved.

Using samples of 100 trials in each parameter combination of

$$a_{min} = 1.7 \quad (0.1) \quad 3.2$$
$$e_{max} = .02 \quad (.02) \quad .20$$
$$\eta = 5, \quad 10$$

and taking $\mathbf{H}(output)/\mathbf{H}(input) = R_H$, by lumping the input and the output frequencies in 10 equal-width categories over the closed ranges occurring, the surfaces of $R_H = f(a, e, \eta)$ may be plotted, as 160 values in each of two $a \times e$ matrices. There is some second-order local irregularity in these surfaces, possibly due to the stochastic nature of the input series samples used in the simulation, but there are trends discernable by inspection such that it is justified to fit, heuristically, a polynomial bivariate linear regression surface as a statistical description, without seeking specific interpretation of the regression coefficients. This has to be done for each η value separately, in the form

$$R_{H:\eta} = b_1 a + b_2 a^2 + b_3 a^3 + b_4 e + b_5 e^2 + b_6 e^3 + \xi \qquad [8.1]$$

and then it is possible to compare the coefficients $\{b_i\}$, $\quad i = 1, 2, 3, ..., 6$ and the distribution of the residuals ξ. It is assumed, from Figures 2.5, 2.6, that powers of a and e beyond 3 are not fruitfully examined.

For $\eta = 5$, the regression surface of R_H on a, e is, omitting terms with negligible contributions,

$$R_H = 1.55 - .15 \pm .08a^3 + 34.29 \pm 14.59e^2 - 106.59 \pm 43.73e^3 \qquad [8.2]$$

with: $s.e.(est) = .078, \qquad R^2 = .626, \qquad r_{Ra} = .603, \qquad r_{Re} = .432,$ $\bar{R}_H = .827 \pm .125$.

For $\eta = 10$, the corresponding surface is rather different;

$$R_H = .69 + 4.93 \pm 1.72e,$$

with: $s.e.(est) = .094, \qquad R^2 = .819, \qquad r_{Ra} = .129, \qquad r_{Re} = .865,$ $\bar{R}_H = .640 \pm .216$.

The results usefully imply that over a range of η values which can generate some recognizable psychophysical functions, not too dissimilar from a CNO, the change in the entropy of the output distribution, for maximum entropy input, depends in $\Gamma \, V \, 1$ both on a_{min} and on e_{max} for brief η, and is moderately predictable. The breakdown in $\mathbf{H}_R$ when induced by some e values creates a bimodal response distribution with high end values of $Y_{obs}(\text{Re})$, irrespective of the actual range of $Y_{obs}(\text{Re})$ values created. This implies that the output information is a partial basis for limiting the region of the system parameter space within which the most probable *a posteriori* values of the parameters lie. The changes in the relation between information transmission and system parameters as the internal loop dwell time η is increased means that to some extent η is identifiable from R_H, because U and $\Delta^1 U$ are known and the mappings $U \mapsto a$ and $\Delta^1 U \mapsto e$ are assumable or partially identifiable from the topology of the gain function. Possibly the identifiability of the mappings is less certain than that of $\eta \mapsto R_H$, for larger η.

Why do Responses stop when a Stimulus Series stops ?

The question now put seems silly, and indeed is silly if psychophysics begins with models which assume no sequential covariance of events from trial to trial, or from moment to moment. As previously noted, in equations [3.3], [3.4], [3.5], sequential independence is not formally introduced into some psychophysical models which have enjoyed wide currency. If, however, results such as those in Table 5.1 closely resemble real data, then there is something to be explained and some novel experiments to be suggested.

A psychological experiment usually consists of a series of trials almost evenly spaced in time; on any one trial J four events can be distinguished (amongst others, such as reinforcement):

B_j a signal that a trial has **B**egun, $= 0$ *or* 1

S_j a **S**timulus with some measurable physical magnitude, U_J

R_J a **R**esponse with some associated quantified measure, Y_J

and T_J a signal to **T**erminate the trial.

In practice, S_J may also serve as B_J, and R_J may also serve as T_J.

A psychophysical experimenter usually seeks to summarise findings in a functional relationship $y = f(U)$, and there may be observable contextual variables $\{X\}$ which are presumed to be constant over the time interval $\tau(T_J, B_J)$, or implicit central (unobservable) variables $\{X'\}$ in the organism which intervene between S_J and R_J. In either case, again an invariant functional relationship $Y = f(X, X', U)$ is sought. In a generalisation of this approach, U_J and Y_J may be replaced by probability measures on sets $\{S\}$ and $\{R\}$. The formal theory of such experiments as a branch of choice theory was developed extensively within the 1960's (Luce, Bush and Galanter, 1963).

From the 19th century onwards (Wundt, 1896, et seq, cited by Boring, 1950) it was clear that subjects can make anticipatory responses to expected events before those events actually occur, so that if we are only concerned to record the occurrence of R_J and not to attach Y_J to it, then the interval $\tau(R_J, B_J)$ can be less than $\tau(S_J, B_J)$ after some previous sequences of trials, $1, 2, ..., j - 2, j - 1$ has happened. The question raised here is, what is the expected distribution of values of Y_J if S_J is null after $S_{J-k}, S_{J-k+1}, ..., S_{J-1}$ is almost always non-null ?

This question at first seems fatuous, because in a typical, and some would say competently conducted, experiment, if no B_{J+1} follows a T_J, and then no S_{J+1} follows B_{J+1}, no R_{J+1} is observed; the subject has got up and walked away. In short, T_J without a subsequent B_{J+1} marks and defines the end of an experiment. This raises no conceptual problems if the representation of the process f is written only in the form $Y_J = f(U_J)$; the current stimulus and only the current stimulus S_J is the cause of the current response R_J. Perseveration in responding, $R_{J+1}, R_{J+2}, ..., R_{J+k}$ after $U_{J+1} = U_{J+2} = ... + U_{J+k} = 0$ can even be taken as a diagnostic sign of very low intelligence or of brain damage.

However, consider the case where (i) B_J is subsumed into S_J, $S(B)$, and (ii) it has been established that

$$Y_J = F(U_J, U_{J-1}, ..., U_{J-h}, Y_{J_1}, Y_{J-2}, ..., Y_{J-g}) \qquad [8.3]$$

where h and g are minimal order estimates by time series analysis; F is usually but not necessarily taken to be linear and stationary in its moments (Gregson, 1983).

To put these conditions another way, the response to a unit input $U_J = 1$ is the impulse response function, over the span of lags 0 to k, on trials $J, J + k$ as

$$I_{J,k} = b + 0 + v_0^{(J)} + v_1^{(J+2)} + ... + v_i^{(J+i)} + ... + v_k^{(J+k)} \qquad [8.4]$$

where the notation $V^{(J)}$ means v at and only at time J. The v are real scalar coefficients. Some of them can be less than or equal to zero, but observed responses are necessarily non-negative. The observed response R_J within a series of mostly non-null S_J (that is, nonzero U_J) is a convolution of $I_{\theta,k}$ within the series $\{U_{J-k}, ..., U_J\}$ from $\theta = J - k$ to $\theta = J$. The coefficients $\{v_i\}$ are estimated from the complete series $\{U, Y(\text{Re})\}$ over $1, ..., J$, that is, under the condition (here denoted by D) that most S_θ are non-null (including S_J) up to $\theta = J$.

To express this in an experimental context in which the $\{U\}$ is a random (independent, identically-distributed, but not necessarily gaussian) series the autocovariance spectrum of $\{Y\}(\text{Re})$ and the cross-covariance spectrum of $\{U, Y\}$ will both be non-null at the lags $1, ..., k$. We need not here consider nonstationarity due to linear trend, as in learning or slow adaptation.

If the process $S \to R$ is stationary and approximately linear this implies that if $U_{J-1} > 0$, $U_J = 0$, then

$$Y_J \quad = b_0 + v_1 U_{J-1} + v_2 U_{J-2} + \dots + v_k U_{J-k}$$
$$Y_{J+1} \quad = b_0 + v_2 U_{J-1} + \dots + v_k U_{J-k+1}$$

$$\vdots$$

[8.5]

$$Y_{J+k} \quad = b_0$$

so an S which is null can also satisfy [8.3] so that $S_{0,J}(B_J)$ elicits $R_J = f(I_{J,k}, U)$ which is by definition non-null under condition D.

Let us rephrase this in a way that leads to alternative predictable experimental outcomes:

H1: If a non-null B_J is a necessary condition for a non-null R_J and $S(B)$ is used, then the S series $U_1, U_2, \dots, U_J, 0, 0, 0, \dots$ will produce an autoregressive $Y(\text{Re})$ series up to $Y_{J,obs}(\text{Re})$ and null responses thereafter.

(This is what is observed in an experiment where the subject only responds when signalled in some way that a trial exists).

H2: If the structure $Y = f(I, U)$ is, with separate non-null B, necessary and sufficient to generate responses, then the S series $U_1, \dots, U_J, 0, 0, \dots$ will produce an autoregressive response series Y which is non-null up to Y_{J+k}.

The time series representation of this H2 situation is neutral with regard to the question of a single $B_J = 1$, zero U_J in the series, as compared with a string of $U_J = 0$ after a $B_{J-1} \neq 0, B_J = 0$. The response to U_J is made by a physically realisable process, and therefore cannot take any account of U_{J+1} at time J.

To put this question in psychological terms, if the observer can be induced to believe that an implicit $B_J = 1$, even if S_J is not detected psychophysiologically, the perceived ΨU_J will be generated by $f(I, U)$ and the response $Y = f(I, U)$ by

$$\Psi U_J = F^{-1}(Y) \qquad [8.6]$$

where F is the transfer function which can be estimated from $\{U, Y\}$ $1, \dots, J$ as in [8.3].

The experimental distinction between a trial in which $U_J > 0$ and $R_J = 0$ as against a trial in which $U_J = 0$ and $\Psi U_J > 0$ and $R_J = 0$ would have, in the above framework, to be taken up by the coefficients of [8.4].

An alternative is to allow a piecewise linear model for $I_{J,k}$ with a threshold below which $\{v_i\}$, the *ltf* coefficients in that subthreshold state, are all zero.

The model [8.3],[8.4],[8.5] makes a series of predictions $\hat{Y}_{J+1|J}$; the distribution of residuals

$$_1R_J = \hat{Y}_{J+1|J} - Y_J \quad | \ U_J > 0$$

and

$$_2E_J = \hat{Y}_{J+1|J} - Y_J \quad | \ U_J = 0 \tag{8.7}$$

are calculable without finding F^{-1}. We may use the corollaries that:

If H1 is true then $_1E_J$ is not distributed as $_2E_J$,

if H2 is true then $_1E_J$ is distributed as $_2E_J$, for up to k trials after the start of a null-S run of trials.

The treatment so far has been concerned with linear autoregressive models with stochastic noise; if we were to assume that f is generated instead by a nonlinear deterministic difference process then the prediction of what will happen is more complicated and perhaps more interesting. In system Γ we have examined a few cases drawn from Γ V 1 and Γ V 7, as shown in Table 8.1 If a block of null-S trials is introduced into a series as an interpolated subsequence, then the output series $Y_{obs}(Re)$ may exhibit a diversity of patterns. It may, for example, drop sharply with a lag to zero (mimicking [8.4]) and recover when $S > 0$ reappears but at a different lag, usually shorter (again mimicking typical [8.3],[8.4] models), or at the other extreme it may perseverate with a negligible diminution of output. What happens depends upon $W, a_{min}, e_{max}, \eta$ and the behaviour of the system under these circumstances, this provides yet another basis for the identification of the system dynamics, complementary to the ones already outlined. That is, only variants and parameter sets which create input-output plots qualitatively resembling the observed drop-offs of responses with the onset of null-S strings can be valid. This stringent criterion eliminates some variants (that is, Ws) entirely; it serves as an additional basis for indentifying system structure apart from tests on the autoregression, the gain function topology of the stimulus-response plots, and information degradation. Some of the Γ simulations are perhaps surprising in their form; cascading can either accentuate or eradicate the effects of local blocks of null-S trials.

Table 8.1

Summary of Simulations on $Y_{obs}(\mathrm{Re})$ only

Cascade $W\Gamma\,V\,1$	lag of onset of drop	period to reach minimum observed	lag of onset of climb	period to restore to previous level
	$a = 2.6,$	$e = .05,$	$\eta = 2$	
C0	0	18	0	17
C1	0	17	1	17
C2	0	16	1	17
	$a = 2.6,$	$e = .15,$	$\eta = 2$	
C0	0	17	0	19
C1	1	15	2	20
C2	4	14	5	20
	$a = 3.4,$	$e = .05,$	$\eta = 2$	
C0	0	17	0	18
C1	14	5	0	5
C2	1	0	0	0
	$a = 2.6,$	$e = .05,$	$\eta = 5$	
C0	0	9	5	17
C1	1	7	14	17
C2	0	7	14	20?
	$a = 3.4,$	$e = .05$	$\eta = 5$	
C0	0	18	3	19
C1	1	14	4	20
C2	7	10	4	10
C6	Indeterminate			
	$a = 3.2,$	$e = .15,$	$\eta = 5$	
C0	0	18	3	18
C1	0	9	3	18
C2	9	7	15	5
C6	9	7	15	5
	$a = 3.2,$	$e = .15$	$\eta = 10$	
C0	0	18	3	20
C1	2	0?	18	8
C2	1	4	18	8

Table 8.1 continued

Cascade WΓ V 7	lag of onset of drop	period to reach minimum observed	lag of onset of climb	period to restore to previous level
	$a = 2.6,$	$e = .05$	$\eta = 5$	
C0	Indeterminate		15	10710?
C1	3? produces a	paradoxical local	rise in output	
	$a = 3.2,$	$e = .15$	$\eta = 5$	
C0	Indeterminate			
C1	10 - 15	3	1	1
	$a = 3.4$	$e = .05$	$\eta = 5$	
C0	Indeterminate			
C1	No detectable	effects		
Γ V 1				
	$a = 2.6$	$e = .05$	$\eta = 10$	
C0	0	20	0	20
C1	Indeterminate			
C2	0	16	27?	0?
	$a = 3.4$	$e = .05$	$\eta = 10$	
C0	2	0	20	2?
C1	2	0	23	0?
C2	9	6	14	6
	$a = 3.2$	$e = .15$	$\eta = 5$	
C0	0	7	2	20
C1	2	8	7	16
C2	0	9	17	7

References

Boring, E. G. (1950) *A History of Experimental Psychology.* New York : Appleton - Century - Crofts.

Gregson, R. A. M. (1983) *Time Series in Psychology.* Hillsdale, New Jersey: L. Erlbaum Associates.

Luce, R. D., Bush, R. R. and Galanter, E. (1963) *Handbook of Mathematical Psychology Vol I .* New York: Wiley.

Michon, J. A. (1967) *Timing in Temporal Tracking.* Soesterberg: Netherlands Institute for Perception RVO-TNO.

Ward, J. A. (1979) Stimulus information and sequential dependencies in magnitude estimation and cross-modal matching. *Journal of Exper-*

imental Psychology: Human Perception and Performance, 5, 444 - 459.

Ward, L. M. (1982) Mixed-modality psychophysical scaling: Sequential dependencies and other properties. *Perception and Psychophysics, 31,* 53 - 62.

9 Matching Data Patterns and Theory Patterns

In traditional psychophysics a statistical model of relatively simple linear form, with attendant gaussian error, is fitted to data by some variant of the method of moments. Refinements, such as the classical Müller-Urban weights (see Woodworth, 1938, chapter XVII) or modern partly Bayesian treatments on model inexactness (Box, 1980) are available, but without it being possible to construct likelihood functions a rigorous treatment of goodness-of-fit questions is generally assumed to be lacking.

Here some possible heuristics on fitting Γ, where a series of variants are *a priori* plausible, to data samples, are briefly suggested with a worked example. The input-output plot $U \mapsto Y(\text{Re})$ may frequently, as has been seen, be treated as a mixture of an ogive and outlying clusters. It looks like that, though it is not generated by adding two such patterns together, but is an integral result of the internal nonlinearities. It is the special capacity of Γ to produce plots that are not simply noisy ogival shapes, but have more complexities, which makes it empirically interesting. However, the price of the obviously non-CNO forms which resemble some actual data is the need to find quite different methods of deciding if data do match theory; the problem becomes one of pattern recognition. As linear psychophysics can predict the CNO part of a distribution, and by strong averaging ignore the peculiarities of the local distribution of outliers or outlier clusters, the part of a data $S - \mathcal{R}$ plot which resembles best a CNO is the least useful in testing a range of Γ variants over a local region of the (a, e, η) parameter space against the nearest linear contenders. What is universally predictable is not discriminative.

No statistical model is a perfect fit to real data, and interest can be with profit sharply focussed on data properties that seem critical for model assessment and revision (Cook, 1986).

The literature on pattern recognition is now very extensive and replete with methods which work only when data properties are narrowly circumscribed. Methods for matching two patterns ('recognition' by the theory of the data) which presuppose absolutely nothing about the form that data will take are notoriously difficult to implement. The K-means algorithm (Tou and Gonzalez, 1974) used here is not very efficient unless the data fall naturally into quite distinct clusters, but it is one of the easiest to program. Modern extensions using Fuzzy Set theory (e.g. Miyamoto and Nakayama, 1986) can be combined with deterministic algorithms to give more flexibility in the recognition of such things as characters or symbols. Here basically we are faced with chains of points which should not naturally be broken into local clusters, and tight clusters, and scattered areas. The chains are almost always associated with a CNO-like arc or a short segment of such a curve.

As Box (1980) observes:

"No statistical model can safely be assumed to be adequate .... checks are always necessary, they may not be sufficient, because some discrepancies may on the one hand be potentially disastrous and on the other be not easily detectable".

He continues:

"to conduct criticism of a model, it is often necessary to estimate provisionally parameters at intermediate stages .... While it is comforting to remember that a good scientific iteration is likely to share the property of a good numerical iteration – that mistakes often are self-correcting, this also implies that (we) must worry particularly about mistakes that are likely not to be self-correcting".

Identification as Matching

There is no such thing as a measure of the appropriateness of a single variant $\Gamma(W, a, \epsilon, \eta)$ to represent a single sample of data; in the approach used here all appropriateness is relative, the meaningful questions centre on finding a parameter set Γ_s in Γ which minimises a mismatch function between theory and data on the salient features which can only be predicted from this specific Γ_s, given that it matches on a feature subset ω which can be modelled by all of a relevant family of $\{\Gamma\}$ based on the neighbourhood in the parameter space of Γ_s.

It follows that the choice between a Γ variant and a linear psychophysical model is based on a comparison of pattern matching on the input -

output $U \mapsto Y_{obs}$(Re) plot, and on those features which are not matched equally well by both models. Comparison of two or more models should include correction for the number of free parameters in the models (Akaike, 1978, Sakamoto *et al* , 1986), unless the models are equivalent in this respect.

K-means Algorithms

Two versions of a pattern identification process. based on the creation of arbitrary clusters, are considered. only the first is used in the experiment chosen here for illustration.

Unmasked Version

Using a program KCLUST.FOR written by MacBride (1987), based on an algorithm of Tou and Gonzalez (1974) which is from the K-means family of algorithms, and incorporating earlier subroutines written by Price in the same laboratory, the stages are as follows. This version was settled on because it is economical in computing time, and yields interpretable results.

1 Given a set of points, which can be either from theoretical or actual data, distributed within a 50 × 50 grid (the arbitrary convention used for generating graphs in this work), one point c_1 is chosen randomly as the first seed to create a cluster C_1. Some point near the middle of the plot may be a sensible choice.

2 Now c_1 is the temporary working centre of cluster C_1, it is not its centroid. The most remote point from c_1 is then found, its distance from c_1 is

$$D_{12} = max(d(c_1, c_2) \; in \; \Re_p^2, \; p = 1 \; or \; 2)$$

and it is assigned to C_2.

3 The distance D_{12} is then used as a basis for determining the first two cluster radii. A fraction f, where $.3 \leq f \leq .4$ is chosen and circles Θ_1, Θ_2 of radius fD_{12} are drawn on centres c_1 and c_2.

4 If no points remain outside these two circles Θ_1 and Θ_2, or if $C_1, ..., C_6$ have been found, then go to step **7**.

5 If two or more points $\{c_3, c_4\}$ exist such that $(c_3, c_4) \ni (\Theta_1, \Theta_2)$, then calculate

$$D_{m:1,2} = max[d(c_3^*, c_1), d(c_4^*, c_2)]$$

where c_3^* satisfies $max(d(c_3, c_1))$ and c_4^* satisfies $max(d(c_4, c_1))$. Pick the c^* from (c_3^*, c_4^*) which satisfies $D_{m:1,2}$ and draw on it as centre a circle Θ_3 with radius

$$\frac{1}{2}f(D_{12} + D_{m:1,2}).$$

6 Repeat step **4**, to search for the next cluster after the ones already found; after incrementing suffices, use the same rules for location and radius of the next cluster.

7 All remaining points, if they exist, are allocated to the circle with the nearest c in $\Re_p^2$.

8 The centres of the clusters are then recalculated in terms of their final membership, to give a weighted set $\hat{C}_1, ..., \hat{C}_6$, where the respective weights are the numbers in each cluster set. The $\{\hat{C}\}$ is either D_c or Γ_v in the proximity routine below.

9 If this method gives fewer than six $\hat{C}$ as the cluster representation of the original scatter of points, then f is adjusted downwards until it does. This step appears not to be usually needed, $f = 4$ generally works.

10 The defects of this method lie in the assumption that discrete clusters are the underlying basis of the total pattern to be summarised. A decision to accept a solution has to be assisted by inspection of any clustering at interim stages.

Masked Version

The Unmasked Version fails to take account of any special status that some part of the total pattern may have in interpreting its correspondence with a psychophysical process. Partly to remedy this, a facility is needed to weight differentially those parts of the pattern which can be critical against those which carry relatively little information about the special features of one alternative model (compare Kempthorne, 1986).

1 Given a family F_m of CNO masks, generated using a NAGFLIB subprogram, which are all of the same shape but moved in the X–axis direction like a sliding template, each of these $f_m \in F_m$ defines a zone within which points may be weighted differently from points outside the zone.

2 The F_m are mapped onto each of the Γ_s variants which are *a priori* near-optimum matches to the empirical data, and then the coordinate subset $\omega_{m,s}$ of Γ_s which is beneath the mask $f_{m,s} \in F_m$, and which also has the maximum masking coverage, is deleted. Call the residual unmasked set $\Gamma_s - \omega_{m,s} = \Gamma_{\omega,s}$.

3 The F_m are mapped onto the empirical data D and the subset ξ_m beneath the mask is deleted. Call this residual set $D_\xi = D - \xi_m$.

4 Compare in turn each of the pairs $(\Gamma_{\omega,s}, D_\xi)$ via the K–means algorithm, as in the Unmasked Version steps **1 - 7**. This method removes the most probable long-chain structure in the input-output plot $U \mapsto Y_{obs}(\text{Re})$ and consequently leaves mainly clusters and scattered unstructured low-density areas.

5 The defect here is that a real data set generated by a CNO plus superimposed heavy gaussian noise may be indistinguishable in a small sample from a Γ variant, if only the $K-$means procedure is used. In such a case, an ancillary procedure could be tried, in that any of the CNO stochastic forms predicts that point density falls off monotonically with distance from the masked region. This trend is not generally a property of all Γ variants, violations of it occur readily with Γ V 1.

Using in turn the $f_m \in F_m$, the density of captured points should drop off in an almost normal fashion along the series

$$F_m, F_{m\pm 1}, F_{m\pm 2}, F_{m\pm 3}, ..., F_{m\pm k}$$

As this property, if violated, constitutes a strong basis for rejecting the CNO stochastic noise model and only masking Γ_s variants, it could be applied if necessary after inspection of data plots, in stage **3** above.

Proximities of Two Clustering Solutions

After applying a clustering algorithm to a given input-output plot, the result is a simplified structure which can be treated as a vector of triples, $\{c_{ix}, c_{iy}, c_{in}\}$, where c_i is a cluster, and x and y are its centroid coordinates, and n is some weighting atttached to it. If necessary, we assume that the set $\{c\}$ is a residual set after masking out the CNO-like features of the plot.

The clustering algorithm in itself simplifies a topology into a form which highlights dense and fairly autonomous local features; whether these are the same features that a human observer would regard as salient depends both on the observer and on the algorithm. The stopping rule within the algorithm, to decide when all and only the important clusters have been isolated, is in this sense crucial. Here it has been assumed that a maximum of 10 clusters could plausibly be given psychophysical meaning; in practice fewer would usually be sufficient. It is worth noting that the SDT analogues (see Chapter 11) of responses into 2×2 tables of true and false detection, and true and false negatives, is a sort of four cluster model. Here, in contradistinction, we set the number of clusters *a priori* after some trial and error at six; they can, off the CNO zone, range in number between zero and at least four in different Γ variants so far simulated. The CNO zone can also break into segments, as for example in Figure 13.4, with two or three parts.

After repeatedly applying the clustering algorithm to the data and to a range of Γ variants, the proximity of the two related clustering algorithms has to be determined. The Γ variant with the highest proximity is the chosen representation. In Bayesian terms, if Γ_v is the model and D_c the data cluster,

$$P(\Gamma_v \mid D_c) \propto P(D_c \mid \Gamma_v) \cdot P(\Gamma_v) \qquad [9.1]$$

and as the number of parameters in the Γ_v is set at 3, (a, ϵ, η), the correction for the system parameters and their associated degrees of freedom (Akaike,

1978) is a constant and so the form [9.1] is equivalent, as a basis for choosing between theories, to an AIC decision.

Using a Fortran program CUPL (Gregson, 1987) with NAGF subroutines,

$$D_c = \{c_{ix}, c_{iy}, c_{in}\}, \quad i = 1, ..., k_d$$

for the data set, and

$$\Gamma_v = \{\gamma_{hx}, \gamma_{hy}, \gamma_{hn}\}, \quad h = 1, ..., k_v$$

for a Γ variant, the problem is to express the proximity of the two cluster sets $\{c\}, \{\gamma\}$. If and only if the data are appropriately clustered and the model is appropriately chosen then $k_d = k_v$, which immediately suggests an iterative process of data clustering via $K-$means, to give $\hat{k}_d$, and model selection to give $\hat{k}_v$, when $\mid \hat{k}_d - \hat{k}_v \mid$ is successively minimised, subject to a proximity measure $\mathbf{P}_{cv}(D_c, \Gamma_v)$ also being minimised.

$\mathbf{P}_{cv}$ is defined on the product set $\{k_d \times k_v\}$; weighted city-block ($\Re_1^2$) and weighted Euclidean ($\Re_2^2$) solutions are candidates given the extensive use of these representations in psychometrics (Gregson, 1975); here $max\Re$ is the biggest distance possible in the square grid plotting the input-output graphs:

$$\mathbf{P}_{cv} \mid \Re_1^2 = \sum_i^{k_d} \sum_h^{k_v} (max\Re_1^2 - \mid c_{ix} - \gamma_{hx} \mid + \mid c_{iy} - \gamma_{hy} \mid) \cdot (c_{in} \cdot \gamma_{hn})^{1/2}) \quad [9.2]$$

and

$$\mathbf{P}_{cv} \mid \Re_2^2 = \sum_i^{k_d} \sum_h^{k_v} (max\Re_2^2 - [(c_{ix} - \gamma_{hx})^2 + (c_{iy} - \gamma_{hy})^2]^{1/2} \cdot (c_{in} \cdot \gamma_{hn})^{1/2})$$

$$[9.3]$$

There exists a subset, size $\min(k_d, k_v)$ of $\mathbf{P}_{cv}$ terms in $\mathbf{P}_{cv} \mid \Re_1^2$ or $\mathbf{P}_{cv} \mid \Re_2^2$ which are associated with the best matches between those clusters which should occur in both data and model. That is, if a set of clusters $\Gamma(k_v)$ exists there should be a corresponding (sub)set $D_\Gamma(k_d)$, $(k_v \leq k_d)$, such that a $k_v \times k_v$ matrix of $\mathbf{P}_{cv}$ terms can be reordered to have maximal terms in its diagonal cells. In geometrical terms, there should exist a transformation of D_c (affine on each axis) which will nearly superimpose $\Gamma(k_v)$ on $D_\Gamma(k_d)$.

As in the example we have fixed $k_d = k_v = 6$, the proximity $\mathbf{P}$ is the sum of the set of *one* $\Leftrightarrow$ *one* cluster distances [9.3] in which the smallest weighted distance (D_1, Γ_x) is found first, then the next smallest $(D_2, \Gamma_{\{v-x\}}) = (D_2, \Gamma_y)$, then $(D_3, \Gamma_{\{v-x-y\}})$, and so on. If the constraint of a *one* $\Leftrightarrow$ *one* is not made, then the closest cluster in Γ may map *many* $\Leftrightarrow$ *one* at some D, which is not wanted.

The subset of k_v greatest proximities is then an overall rough relative measure of the appropriateness of the best Γ variant. The trace $\mathbf{T}_v$ of the $k_v \times k_v$ matrix is an appropriate index.

An alternative approach is to create the full matrix of $k_v + k_d \times k_r + k_d$ proximities, which has both the within- and between-solutions inter-cluster measures like [9.2] and [9.3]; that is, two subdiagonal triangular matrices for D_c and Γ_v, and the $k_v \times k_d$ rectangular matrix between D_c and Γ_v. The total matrix is symmetric, and real, positive in all cells, so its eigenvalues are all positive. The eigenvectors yield information on the pairing of clusters within and between solutions.

Experiment

Method

In an olfactory psychophysical study on the perception of suprathreshold mixtures of Bergamot (Givaudan) and n-Butanol, series of 55 trials were presented with trials spaced at 30 second intervals. The mixtures ranged in equally spaced steps from 100% Bergamot + 0% n-Butanol, through 90% Bergamot + 10% n-Butanol, 80% Bergamot + 20% n-Butanol,......, to 0% Bergamot + 100% n-Butanol, so that the series is approximately one of almost constant total intensity (T) mixtures, and the concentrations and approximately the perceived intensities of the two components vary inversely. The presentation series were generated randomly in an initial series of 11, and thereafter cyclically permuted to give a total of 55 trials. The theoretical rational for being interested in performance in this task arises in a study (Gregson, 1980) on paradoxical mixture perception where marked nonlinearities are expected. Details of apparatus and other relevant methodological details are given in Gregson (1984b, 1986) and Gregson and Gates (1985), and are analogous to the study described later in Chapter 12. The data used here are only part of a larger experiment and are selected to illustrate the use of the proximity measures in matching theory to data.

The task on each trial was for the subject to rate, as a number, the intensity of the odour associated with Bergamot (B) and the total perceived intensity (T), in that order, after having initially been given, before the test series, a reference stimulus of 100% Bergamot, at a concentration of $950ml/min$ in $150l/min$ of filtered air, designated by the experimenter as "100 B units". The subsequent mixture components could be rated higher than 100 if the subject so wished. Stimuli were presented for 3 seconds at regular half minute intervals.

In this illustrative analysis, only the relation between Bergamot vapour concentration (as S and U) and B responses (as $\mathcal{R}$ and Y_{obs}) is examined, to see if some matchings of stimulus-response plots to Γ V 1 plots can be established using the algorithms outlined in this chapter.

Table 9.1
Likelihood Ratios of Proximity Measures $\mathbf{P}_{cv}$ [9.3]
for Real Data compared with Random Simulations
averaged over a set of Γ V 1 models from $0 < e < .3$

From	To	Λ
0	575	Nil
576	600	5.49
601	625	1.13
626	650	27.20
651	675	160.97
676	700	494.24
701	725	817.76
726	750	241.49
751	775	319.15
776	800+	Nil

'Nil' means no cases existed.

The purpose of Table 9.1 is to show that random distributions of clusters are distinguishable from the patterns produced by actual data, when compared against theory.

Results

In a complicated task of this sort, subjects do not necessarily behave homogeneously; some perform relatively accurately and some others are apparently induced by the overwhelming complexity of the stimulus presentation conditions to respond almost paradoxically, that is, they either show negligible sensitivity to odourant component concentration differences in the simple $S - \mathcal{R}$ plot, or they may even respond inversely, so that perceived intensity decreases with increasing component concentration. It is as though they confuse components and then respond to the wrong one, but it is not such a confusion which necessarily underlies their observed behaviour. The way that the present task is constructed of course maximises the probability of eliciting such behaviour if it is going to occur at all.

Eighteen subjects served in this experiment, and each was analysed separately. To investigate the internal dynamics of responding the $S - \mathcal{R}$ sequences were treated as binary time series and analysed using an SCA computer package to establish the form and extent of sequential dependencies in the data. Figures 9.1 and 9.2 show examples of a data plots and theory for the qualitative features to be compared in this example.

The model fitted here, as a common screening filter on all data sets,

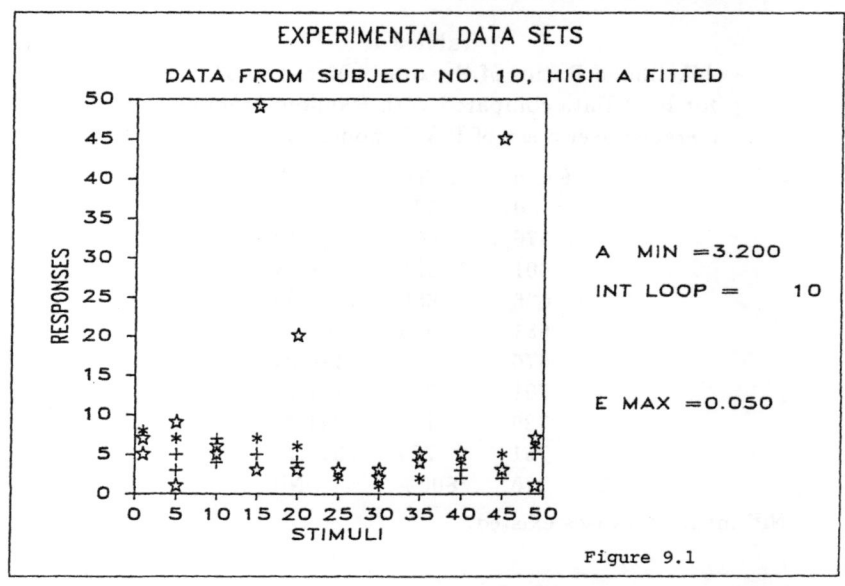

Figure 9.1

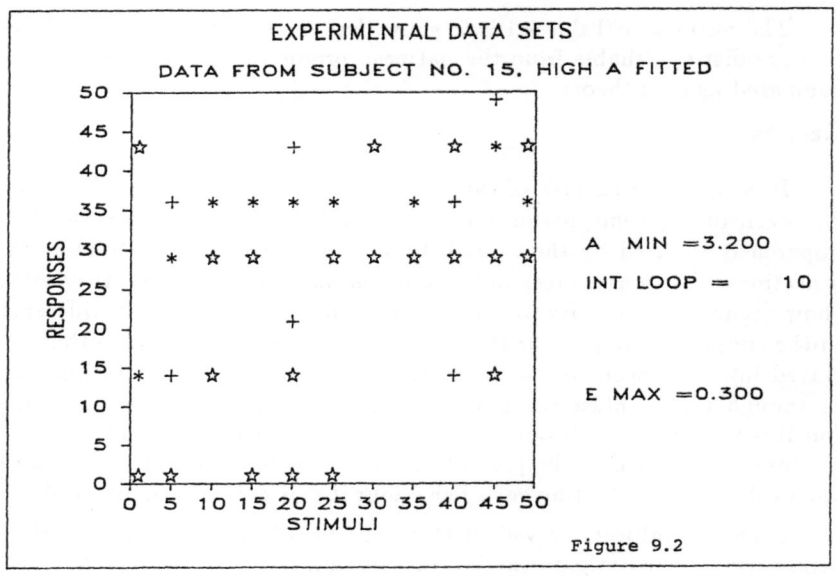

Figure 9.2

is of the form

$$y_j = \frac{B(z^{-1})}{A(z^{-1})} \cdot u_j + \frac{D(z^{-1})}{C(z^{-1})} \cdot e_j \qquad [9.4]$$

where z is a shift operator over one intertrial interval. z^{-1} notation is fairly standard in control theory and here we are consistent with the usage of Young (1984), (compare the analogous approach in Chapter 12), so

$$x_j = \frac{B(z^{-1})}{A(z^{-1})} \cdot u_j \qquad [9.5]$$

is the system model,

$$\varepsilon_j = \frac{D(z^{-1})}{C(z^{-1})} \cdot e_j \qquad [9.6]$$

is the noise model, and hence

$$y_j = x_j + \varepsilon_j. \qquad [9.7]$$

To sustain major details of the analysis used here in Table 9.2 it is sufficient to write

$$B(z^{-1}) = b_0 + b_1 z^{-1} \qquad [9.9]$$

though terms up to z^{-9} were used in calculating the transfer function co-efficients which underly the factorization parameters σ, $i\omega$ set out in Table 9.2. Only coefficients b_0 and b_1 are needed to point up some different re-sponse modes into which subjects here fell. Our interest focusses on seeing how well some parameter sets in Γ V 1 can fit the observed $S - R$ topology, and if differences in values of a or e are related to the internal dynamics of the sensory process as identified by the time series modelling. No for-mal hypotheses are advanced at this point; the analysis is exploratory and illustrative of the use of the algorithms described in this chapter.

Table 9.2 contains the following information;

Subject reference number; this was the order in which subjects partic-ipated, it is given to illustrate that no trends over time in the conduct of the experiment, due to environmental factors, are discernable.

reg % refers to the variance taken up by a linear regression on the raw data plots of $S - R$, it is generally negligible.

filt % is the covariance of residuals of input and output after both have been filtered by a purely autoregressive model fitted to the stimulus series (Bergamot values) after differencing once to remove trend due to slow adaptation. This is a usual so-called "prewhitening" for time series modelling.

b_0 and b_1 are estimated coefficients in equation [9.9], where the esti-mates are greater than 1.2.

R^2 is the multiple correlation coefficient for the regression of the fitted model on the filtered input for the full model [9.4] used, with orders for $B(z^{-1}) = 7$, $A(z^{-1}) = 3$, $D(z^{-1}) = 3$, $C(z^{-1}) = 3$. It is seen that R^2 can be over 300 times greater than the regression % variance of the linear regression fitted to the raw $S - R$ plot. The d.f. for R^2 is not less than 28. It may here be interpreted as a measure of the extent to which the roots of the factorized transfer function based on [9.4] are likely to be a good representation of the system dynamics.

σ and $i\omega$ are the coordinates in w-space of the largest conjugate (complex) pair of roots, which furnishes information on the rate of convergence (in the left-hand, negative, half-plane) of the system dynamics to stability after an impulse input, and of the amplitude of the oscillations associated with that root. See Gregson (1984a) for a detailed pictorial illustration of this interpretation, and references by Gregson (1983) to w-space dynamic representation. The w transformation is also given in Churchill (1960, Section 41, and page 286 Fig. 12) as the *linear fractional transformation*.

Table 9.2

Some parameters of Fitting Regression and Time Series Model [9.4] to Data

S #	reg %	filt %	b_0	b_1	R^2	σ	$i\omega$
1	2.29	1.44	-	-	.935	-.76	1.66
5	1.59	9.00	-	-	.613	-.17	1.28
6	1.25	0.49	-	1.95	.927	-.18	1.15
9	2.25	8.41	2.18	3.38	.903	-1.11	2.06
10	0.26	6.25	-5.29	-2.65	.979	-0.08	1.16
12	7.67	9.00	-4.12	-	.923	-0.02	1.13
14	1.06	8.41	2.62	-	.824	0.07	1.18
15	11.22	0.25	-	5.95	.881	-0.26	1.52
—	—	—	—	—	—	—	—
2	14.89	2.25	-	1.59	.841	-1.02	1.30
3	1.24	0.49	-	1.54	.885	-0.18	1.84
4	7.23	1.96	-	-	.661	.53	1.63
7	0.29	1.44	-	-	.876	-0.93	2.67
8	0.01	0.49	-	-	.851	-0.09	2.68
11	0.07	0.09	-	-2.54	.825	-0.78	1.99
13	0.40	0.36	-	-2.50	.934	0.13	2.68
16	2.03	7.29	1.27	5.42	.959	0.01	1.68
17	4.21	0.09	-2.31	2.06	.824	0.12	1.86
18	4.85	0.01	-2.09	-	.678	-0.09	1.79

If the σ (real) and $i\omega$ (imaginary) parts of these roots are rewritten in polar coordinates, M, Θ, where M is the root modulus and Θ is the angle from the origin measured anticlockwise from the positive σ axis, the M reflects the dominance of the dynamic component in the system and Θ the rate of convergence to zero. The trajectory of the loci of these roots in a system with the same feedback but varying sensitivity may be plotted, and for the cases above the dashed line in Tables 9.2 and 9.3 is almost a straight line, slope = -.78, intercept = 1.15, F = 62.59, (d.f. = 1,6 $p < .001$). This is illustrated within Figure 9.3.

The eight cases above the dashed line in the Tables 9.2 and 9.3 all had $Z \geq 3$ for $\mathbf{P}_{cv}$ in $\Re_2^2$ when fitting Γ models; the interest here is that these cases are associated with relatively stable dynamics and lie on a definable trajectory in the w-space. Some of the remaining 10 cases have root loci which suggest they are on or near to limit cycling and thus transmit relatively little stimulus information. We should note carefully that the limit cycling detected here is in the input-output of the system, modellable if at all in $U \mapsto Y_{obs}$ and not the limit cycling that arises in the internal recursive loop from the dynamics of [2.2]. Conditions under which the latter can generate the former have not been indentified yet.

Thus, the cases where the Γ V 1 (or V 7) model fits relatively well, in the sense of the proximity algorithm used here, are also those cases which are dynamically relatively efficient, and in all but one case (# 15) are associated with $a_{min} = 3.2, .05 \leq e \leq .10, \eta = 10$. There may exist other Γ variants which fit better; this example is not in this sense exhaustive and has only explored a very limited range of the parameter space $\{a, e, \eta\}$, but the methodology for comparing the goodness-of-fit of other variants would be the same.

The Γ V 1 with $\eta = 10$ cannot, over the range of $\{a_{min}, e\}$ used here explored ever produce conjugate roots in the Y_{obs} series near to the $i\omega$ axis outside the unit circle on the origin. But if η is increased to 20, to represent a relatively sluggish or perseverative internal loop dynamics, then it is possible to produce roots in the area representing limit cycling in the output. The model will also fit reasonably in terms of the proximity measure $\mathbf{P}$. This is interesting, because it predicts an interpretable form of individual differences in the internal psychophysics which links together the observed input-output behaviour and the sequential properties of responses. The link follows naturally from the properties of the Γ V 1 model without the need to make additional assumptions involving other parameters or processes coupled in series with it.

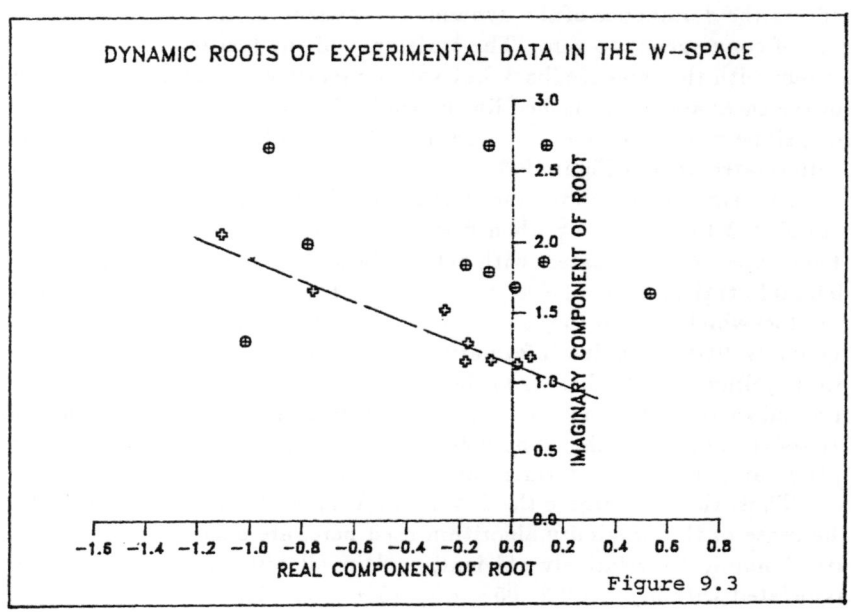

Figure 9.3

References

Akaike, H. (1978) Comments on 'On Model Structure Testing in System Identification'. *International Journal of Control, 27,* 323 - 324.

Bailey, T., and Cowles, J. (1984) Cluster Definition by the Optimization of Simple Measures. *IEEE Transactions on Pattern Analysis and Machine Intelligence, PAMI-6,* 645 - 652.

Ben-Bassat, M. (1982) Use of Distance Measures, Information Measures and Error Bounds in Feature Evaluation. *In* Krishnaiah, P. R. and Kanal, L.N. (Eds.) *Handbook of Statistics, Vol. 2,* 773 - 791. Amsterdam: North-Holland Publishing Co.

Box, G. E. P. (1980) Sampling and Bayes' Inference in Scientific Modelling and Robustness. *Journal of the Royal Statistical Society, Series A, 143,* 383 - 430.

Churchill, R. V. (1960) *Complex Variables and Applications.* New York: McGraw Hill.

Cook, R. D. (1986) Assessment of Local Influence. *Journal of the Royal Statistical Society, Series B, 48,* 133 - 169.

Dubois, S. R., and Glanz, F. H. (1986) An Autoregressive Model Approach to Two-dimensional Shape Classification. *IEEE Transactions on Pat-*

Table 9.3

Maximum Normalized Deviation (Z) of Γ fitted
to data by K-means type algorithm, with associated e,
for $a = 3.2$, $\eta = 10$, in either Γ V 1 or V 7 models.
The Z comparison is with random data as in Table 9.1
but calculated separately for each parameter set $\{a, e\}$

	Γ	V	1	Γ	V	7
S #	P_{cv}	max Z	e	P_{cv}	max Z	e
1	720	2.7	.10	695	0.46	.30
5	730	6.1	.05	611	-2.54	.30
6	740	6.4	.05	627	-1.96	.30
9	720	5.6	.05	612	-2.50	.30
10	780	7.7	.05	638	-1.57	.30
12	700	4.7	.05	644	-1.14	.10
14	700	4.9	.05	587	-3.73	.10
15	760	3.0	.30	701	1.45	.10
—	—	—	—	—	—	—
2	690	1.6	.10	670	0.05	.10
3	630	2.3	.05	716	1.21	.30
4	680	1.3	.10	705	1.64	.10
7	610	1.8	.05	694	0.43	.30
8	650	0.4	.15	663	-0.27	.10
11	690	1.1	.30	682	0.59	.10
13	620	2.2	.05	693	1.04	.10
16	610	1.7	.05	721	1.39	.30
17	660	0.8	.15	688	0.86	.10
18	700	1.7	.30	699	1.36	.10

tern Analysis and Machine Intelligence, PAMI-8, 55 -66.

Fischler, M. A. and Bolles, R. C. (1986) Perceptual Organization and Curve Partitioning. *IEEE Transactions on Pattern Analysis and Machine Intelligence, PAMI-8* , 100 - 105.

Gregson, R. A. M. (1975) *Psychometrics of Similarity.* New York: Academic Press.

Gregson, R. A. M. (1980) A model of paradoxical odour mixture perception. *Chemical Senses and Flavor, 5,* 257 - 269.

Gregson, R. A. M. (1983) *Time Series in Psychology.* New Jersey: L. Erlbaum Associates.

Gregson, R. A. M. (1984a) Invariance in Time Series Representations of 2-input 2-output Psychophysical Experiments. *British Journal of Mathematical and Statistical Psychology, 37,* 100 - 121.

Gregson, R. A. M. (1984b) Similarities between odour mixtures with known components. *Perception and Psychophysics, 35,* 33 - 40.

Gregson, R. A. M. (1986) Qualitative and aqualitative intensity components of odour mixtures. *Chemical Senses, 11,* 455 - 470.

Gregson, R. A. M. and Gates A. (1985) Cross-modal identification: effects of contingent changes in the stimulus series. *Biological Cybernetics, 52,* 247 - 258.

Keller, J. M., Gray, M. R. and Givens, J. A. Jr. (1985) A Fuzzy *K*- Nearest Neighbour Algorithm. *IEEE Transactions on Systems, Man, and Cybernetics, SMC-15,* 580 - 585.

Kempthorne, P. J. (1986) Decision-theoretic Measures of Influence in Regression. *Journal of the Royal Statistical Society, Series B, 48,* 370 - 378.

Miyamoto, S. and Nakayama, K. (1986) Similarity Measures Based on a Fuzzy Set Model and Application to Hierarchical Clustering. *IEEE Transactions on Systems, Man, and Cybernetics, SMC-16,* 479 - 486.

Pavlidis, T. (1977) *Structural Pattern Recognition.* Berlin: Springer-Verlag.

Sakamoto, Y., Ishiguro, M. and Kitagawe, G. (1986) *Akaike Information Criterion Statistics.* Amsterdam: D. Reidel.

Tou, J. T. and Gonzalez, R. C. (1974) *Pattern Recognition Principles.* London: Addison-Wesley.

Woodworth, R. S. (1938) *Experimental Psychology.* New York: Henry Holt.

Young, P. (1984) *Recursive Estimation and Tine-Series Analysis.* Berlin: Springer-Verlag.

10 Metric or Nonmetric Scaling: Properties of Outputs

The input to some Γ V x is by definition a set of real positive numbers U in the interval $(0,1)$. As any number in the interval can be an input value except where some special distribution is created, for example in Chapter 11 as used for simulating an analogue of SDT, and as

$$U = m_{\mathcal{S}}(\mathcal{S}), \qquad [10.1]$$

$$m_{\mathcal{S}}(\mathcal{S}_1) + m_{\mathcal{S}}(\mathcal{S}_2) = m_{\mathcal{S}}(\mathcal{S}_1 \odot \mathcal{S}_2) \qquad [10.2]$$

then it should be possible to treat U as satisfying the axioms of an interval scale m, though only within the bounded range $(0,1)$. There are special problems associated with such bounded metrics, and possible solutions can be based on a $tanh^{-1}$ transformation of the bounded set of values (Schönemann, 1983, Borg and Staufenbiel, 1986). The axioms which necessarily and sufficiently define an interval scale are given by Pfanzagl (1968) and are used here as a starting point.

The problem to be examined is: given that $\{U\}$ is acceptable as on an interval scale, will the corresponding output values $\{Y_{obs}(\mathrm{Re})\}$ also satisfy properties which enable them to be accepted as a case of an interval scale based on operations ? To parallel [10.1], if $U = m_{\mathcal{S}}(\mathcal{S})$ is interval measurement, then what is $m_{\mathcal{R}}$ in

$$Y_{obs}(\mathrm{Re}) = m_{\mathcal{R}}(\mathcal{R}) \ ? \qquad [10.3]$$

Suppose initially that the observations of both U and $Y(\mathrm{Re})$ are error-free; that is, there is zero observational tolerance on any outcome of a single bisection operation in the form of some U_J, $Y_{obs,J}$. In theory, as Γ is deterministic, this is so. In practice, there can be errors of measurement on any U_J or $Y_{obs,J}$, but more usually it is the custom to treat Y as the more imprecise observation. For example, in calculating the regression of Y on U or $logY$ on $logU$ in a magnitude estimation psychophysical experiment, it is treated as a problem in minimising $\hat{Y}$ errors given U.

Pfanzagl showed that a test on a set Ω of numerical observations of data, to see if that set constitutes a basis for an interval scale $\{\odot, \Omega\}$, where $\odot$ is an operation of comparison, is provided by
$\forall \alpha_i, \alpha_{i'}$, where $|$ is the bisection operation,

$$\alpha_i \mid \alpha_{i'} = [\alpha_i \mid (\alpha_i \mid \alpha_{i'})] \mid [\alpha_{i'} \mid (\alpha_i \mid \alpha_{i'})] \qquad [10.4]$$

which means, in experimental operations, that we should

(i) First find, given two responses $Y_{\alpha:i}, Y_{\alpha:i'}$, the stimulus $U_{i,i'}$ which corresponds to the response $(Y_i + Y_{i'})/2$, for convenience denoted by $Y_{i,i'}$.

(ii) Then find the stimulus $U_{i|(i|i')}$ which corresponds to the response $(Y_i + Y_{i,i'})/2$; this is $Y_{i|(i|i')}$.

(iii) Similarly find $Y_{i'|(i|i')}$ and hence $U_{i'|(i|i')}$.

(iv) Now find the stimulus which corresponds to the bisection judgment $Y_{i|(i|i')} \mid Y_{i'|(i|i')}$.

(v)

$$\text{Iff} \quad Y_{i|(i|i')} \mid Y_{i'|(i|i')} = Y_{i,i'} \qquad [10.5]$$

then

$$U_{i,i'} = U_{i|(i|i'))|(i'|(i|i')) } \qquad [10.6]$$

A test for a scale that is interval can thus be constructed by comparing for all, or for a large fraction of pairs, (i, i'), to see if [10.6] holds. The existence of an interval scale is given by the axioms which $\{\odot, \Omega\}$ should satisfy;

Ax10.1	$\odot$ is bisymmetric
Ax10.2	$\odot$ is cancellable
Ax10.3	$\odot$ is continuous
Ax10.4	$\odot$ is connected.

The bisymmetry condition **Ax10.1** entails $\forall \alpha_i, \alpha_{i'}$

$$(\alpha_1 \odot \alpha_2) \odot (\alpha_3 \odot \alpha_4) = (\alpha_1 \odot \alpha_3) \odot (\alpha_2 \odot \alpha_4) \qquad [10.7]$$

Condition [10.4] is a necessary and sufficient test for **Ax10.1**.

The cancellability condition states that

$$\forall \alpha_i \text{ If } (\alpha_1 \odot \alpha_2) = \alpha_4$$

and if

$$(\alpha_1 \odot \alpha_3) = \alpha_4$$

then

$$\alpha_2 = \alpha_3 \qquad [10.8]$$

These two axioms can be tested on a finite sample Ω_0; they are, in principle, falsifiable for $\Omega_O \subset \Omega$ (Pfanzagl, 1968, p. 107). But **Ax10.3** and **Ax10.4** can only be falsified globally on an infinite sample, they are sometimes therefore called "technical" axioms.

In this examination of some samples from Γ V 1 and V 7 the focus is solely on [10.6], for if this is not true then it is not proper to proceed further in that variant to search for evidence that psychophysical responses are measureable up to an interval scale.

An immediate problem arises, which is one even more exacerbated in experimental data; namely, when do we accept that two stimuli or two responses are equivalent (and may be then given equal numerical representation), if it is known that observational errors of unknown degree are attached to our estimates of U and Y ? The axioms **Ax10.1** to **Ax10.4**, and [10.5] - [10.7] are all associated with an ideal case in which the observed events are precise and error-free. Experiments which attempt to check [10.5] - [10.7] have to adopt some inbuilt convention to decide how big is a 'tolerably small' inequality in [10.6] (Raslear, 1982).

It is being assumed here, as is more usual[1] that an observer can given $Y_{\alpha:1}$, $Y_{\alpha:2}$, ($Y_{\alpha:2} > Y_{\alpha:1}$), select a $Y_{\alpha:1,2}$ such that

$$Y_{\alpha:2} - Y_{\alpha:1,2} = Y_{\alpha:1,2} - Y_{\alpha:1} \qquad [10.9]$$

and the problem then is to find $U_{\alpha:1,2}$ corresponding to $Y_{\alpha:1,2}$, *if such a U exists* .

If the starting point is Fechnerian psychophysics or the Plateau - Stevens power law [3.2], then both U and Y are assumed to satisfy **Ax10.3**, and given any interval $\langle U_{\alpha:1}, U_{\alpha:2} \rangle$ we may fill this interval with some U (which is so here) and the Y are considered to be analogously continuous. Unfortunately, this is not the case for all variants of Γ, and to illustrate this in both intractable and tractable cases some alternative parameter sets for Γ V 1 and Γ V 7 are here examined. We are moving towards the position that for some $\{a, e, \eta, W\}$ the Y observations would be expected to be metric, and for others they would not; in a nonlinear psychophysical system there is no such thing as a universally metric output. A universal psychophysical function is a will-o'-the-wisp. As the voluminous literature

---

[1] An exception is Eisler (1980), who noted that both rats and humans could make similarity responses (we hesitate to write of judgments in rats) and for a model of similarity he assumed then showed that bisection points would be geometric means of stimulus pairs.

of experimental psychophysics contains (i) evidence of data which are good approximations to an interval scaling, (ii) cases which have been unanalyzed and are indeterminate, and (iii) cases which obviously violate all of **Ax10.1 - Ax10.4**, this tentative conclusion is not to be thought disheartening. Rather, it is of interest to see if it is possible to identify conditions in $\{a, e, \eta, W\}$ which appear to induce a nearly metric scale output. By "nearly metric" is meant not detectably nonmetric over some closed and unbroken range of response values.

If observations of the response $\hat{Y}$ are imprecise then errors can arise concerning the metric properties of a set of such $\hat{Y}$. In [10.9], suppose that having found a $Y_{\alpha:1,2}$ (a response that satisfies the task of bisecting the perceived interval $m\mathcal{R}(\mathcal{R}_{\alpha:1} - \mathcal{R}_{\alpha:2})$), the data set $\{U, Y_{obs}\}$ is then searched for a $U_{\alpha:1,2}$ which corresponds uniquely with $Y_{\alpha:1,2}$. That is,

$$Y_{\alpha:1,2} = \mathbf{\Gamma}(U_{\alpha:1,2} \mid a, e, \eta, W) \qquad [10.10]$$

so

$$U_{\alpha:1,2} = \mathbf{\Gamma}^{-1}(Y_{\alpha:1,2} \mid a, e, \eta, W). \qquad [10.11]$$

Here the symbol Γ is used as a generic label for the family of transformations in Chapters 5 and 6.

If in [10.11] $\mathbf{\Gamma}^{-1}$ is *one* $\leftrightarrow$ *one* then the continuity assumptions about U lead us to expect that a solution will always exist, that is when $\mathbf{\Gamma}(* \mid a, e, \eta, W)$ satisifies **Ax10.3** and **Ax10.4** then it is generally safe to proceed to look for $U_{\alpha:1,2}$ given the existence of any $Y_{\alpha:1,2}$. Clearly, if the $U - Y$ plot is not almost a continuous single-valued function $Y_{obs}(\text{Re}) = f(U)$ then complications ensue. It is helpful to examine an interestingly bad case, to highlight the problems that can arise.

Suppose, having found $Y_{\alpha:1,2}$ a search is made for a corresponding $U_{\alpha:1,2}$. In the data generating the bisection problem $(Y_{\alpha:1}, Y_{\alpha:2}, U_{\alpha:1}, U_{\alpha:2})$ it is not necessarily the case that there exists one and only one $U_{\alpha:1,2}$ given that $Y_{\alpha:1,2}$ can be obtained in a real experiment or calculated as an arithmetic mean $(Y_{\alpha:1} + Y_{\alpha:2})/2$ in simulation. Searching the finite set of $\{U\}$ used in a simulation (here of 100 values $\sim rect(z)$) of Γ V 1 , in the first instance graphically, as in Figures 10.1 and 10.2, suggests that this pattern is something like part of the upper half of a CNO *and* a scatter of isolated points below; resembling a branch of a tree scattering leaves in a light wind. It also resembles a psychophysical experiment in which on a large proportion of trials there is spasmodic failure to estimate accurately the intensity of stimulation, and consequently stimuli are falsely judged to be null or to be very weak. This happens readily in the chemical senses when stimuli near the absolute detection threshold region are used, a fact which partly motivated this analysis. It follows in these circumstances that for any U there can be a multiplicity of Y values. In the stochastic treatment of psychophysical theory one might then average $\bar{Y} \mid U$ and call this $\bar{Y}$ the estimated response $\hat{Y} \mid U$.

The problem becomes odder if we observe that there are gaps in the U values so that no U exactly matches

$$U_{\alpha:1,2} = \mathbf{\Gamma}^{-1}(Y_{\alpha:1,2} \mid a, e, \eta, W) \qquad [10.12]$$

If it is assumed that $Y_{\alpha:1,2}$ is error-loaded, then depending upon the magnitude of $\pm \delta U$, the observation error, there are consequent limits on U. Computationally, if $U_{\alpha:1,2}$ does not exist but some U_α within the range $U_{\alpha:1,2} \pm \delta U$ does exist as $U^*_{\alpha:1,2}$, then $U^*_{\alpha:1,2}$ is taken as the bisection point. The problem is, that with increasing δU the possibility of identifying what sort of metric $m_{/\mathcal{R}}$ is, is seriously diminished. There are two sorts of error implicit here, errors in observed U and in observed Y; the Γ system is deterministic and so the tolerance δU has rather a different status, as a decision criterion, compared with estimated residual observational error in a stochastic model. Raslear (1982) has drawn attention to the Theory of Error (Beers, 1957) in this context and comments (p.96) "unless the precision and/or accuracy of psychophysical measurement is specified the discussion of substantive issues...may be premature". This means that tests on the compatibility of Y with an interval scale could depend on the choice of δU; if δU is zero then it appears that Γ V 1 could never generate interval response scales over a wide range of its parameter space. But this comment anticipates a little.

Testing the Condition [10.4]

Given a data set, which for simulation is on 100 trials of Γ V 1, and using a program METRIC.FOR on a DEC20,

1 The U series in generated randomly, U_J, $J = 1, 2, ..., 100$, and any U_J takes an integer value $0 < U_J < 50$. From these every tenth U is taken to give $U_{\alpha i}$, $i = 1, .., 10$ as a subset of instances to generate the $10 \cdot (10 - 1)/2 = 45$ pairs upon which bisection is performed.

2 Γ V 1, with $a = 3.2, e = .3, \eta = 10$ generates a corresponding set $Y_J(\text{Re})$. The Im parts are not used here.

3 The bisection of $Y_{\alpha:i}$ values is made, to give the lower triangular matrix $\mathbf{Y}_{\alpha:i,i'}$, $i = 2, .., 10$, $i' = 1, .., i$.

4 A test is made to see if there exists a $U_{\alpha:1,1'}$ for each $Y_{\alpha:i,i'}$ generated in step 3. This is affirmed if there exists a U_J such that $U_j - \delta U < U_{\alpha:i,i'} < U_J + \delta U$.

5 A range of δU values is explored; .25, .5, 1.0, 2.0. This δU represents tolerance due to suspected imprecision of measurement on Y or on U; in this case the treatment of $\{U_J\}$ as a set of integer values instead of taking any real value in the range 0,50 (which is mapped by a scalar multiplier onto the range $a_{min} < a < a_{max}$) is a convention which means that the $U_{\alpha:i,i'}$ can only be selected from a range of integers; if the psychophysical experiment being simulated rests on the classical Method of Production. Setting $\delta U > 0$ is analogous to giving instructions to a hypothetical subject to "pick the stimulus from the range given which is nearest to one that bisects the range between the two reference stimuli with which you began".

6 We remark that it also follows that such a method, unless the stimuli are separated by physical steps less than a *jnd*, is probably incapable of testing if an interval scale exists even locally. Manders (1981, p. 228) notes that if we require that representations are homomorphisms *onto* a numerical structure N, then "we appear to be requiring further relationships among the measured values and the numerical domain, e.g. that any number could be a measurement value". Here the lesser requirement of homomorphisms *into* N is accepted.

7 Because $U \Rightarrow Y$ is *one* $\Rightarrow$ *many*, and $Y \Rightarrow U$ is *one* $\Rightarrow$ *many* in some cases of $\Gamma(a, e, \eta, W)$ there can be a number of different $U_{\alpha:i,i'}$ for any $Y_{\alpha:i,i'}$. So, multiple solutions to stage **5** can exist, the number differing with each bisection. For computational purposes, not more than 10 such solutions have been used for any one (i, i') pair.

8 It is also possible, depending on the value chosen for δU, that no U solution for a given $Y_{\alpha:i,i'}$ exists. Only if δU is increased will there eventually be trapped a $U^*_{\alpha:i,i'}$ which satisfies [10.12].

9 The set of computed solutions to $\forall i, i'$, $U_{\alpha:i,i'}, Y_{\alpha:i,i'}$ are stored and used in the next step, to calculate terms $Y_{\alpha:i} \mid Y_{\alpha:i,i'}$ and $Y_{\alpha:i'} \mid Y_{\alpha:i,i'}$. Again, these bisection comparisons admit of multiple solutions in U for each $Y_{\alpha:i,i'}$ used. The process with $\delta U > 0$ yields a tree structure, proliferating solutions with each successive bisection-of-bisections generation.

10 The comparison of the U associated with $Y_{\alpha:i,i'}$ and $Y_{\alpha:i} \mid Y_{\alpha:i,i'} \mid Y_{\alpha:i'} \mid Y_{\alpha:i,i'}$ is by [10.5] a test of the interval axioms; call the resultant inequality $R_{\delta U}$. This $R_{\delta U}$ should be zero and if some tolerance is admitted, it will have a distribution of values less than δU. The distribution of $R_{\delta U}$ varies with δU.

11 The form of the frequency distribution of $R_{\delta U}$ for 45 comparisons, each with up to 10 solutions, is given in Tables 10.1 and 10.2. $R_{\delta U}$ values beyond the tabulated range are not listed.

It is apparent that the output of Γ V 1 is not compatible with an interval scale, a deduction which could easily have been arrived at just by looking at the specimen plots of Figures 10.1 and 10.2. The point of this analysis is to set up a framework for comparing a diversity of simulations and data, not only for looking at obviously strange cases.

The case of Γ V 7 is rather different; under some circumstances as

Table 10.1

Frequency distributions of R_U for Γ V 1
with $a = 3.4, e = .3, \eta = 10, N = 100$

δU units	.25	.5	1.0	2.0
0	0	40	0	6
1	0	45	5	16
2	0	0	0	56
3	0	4	5	15
4	0	0	0	0
5	0	0	6	24
6	0	0	38	0
7	20	20	18	4
8	0	0	26	8
9	10	1	0	0

Table 10.2

Frequency distributions of R_U for Γ V 1
with $a = 3.2, e = .25, \eta = 10, N = 100$

δU units	.25	.5	1.0	2.0
0	5	45	0	3
1	0	5	10	0
2	0	10	4	2
3	0	0	0	7
4	0	0	14	0
5	0	0	0	2
6	0	0	0	4
7	0	2	38	38
8	0	0	0	5
9	0	0	10	0

shown in Table 10.3 it yields a well-behaved tight CNO-like plot, which for all but very small δU would give an $R_{\delta U}$ distribution lumped near to zero, as the error-free CNO is a $\Gamma \pm \delta U \leftrightarrow 1$ homomorphism of U (interval) onto Y. See Figures 10.3 and 10.4.

There are no a priori indifference relations on U, but using any

$$\langle U - \delta U, U + \delta U \rangle = U^* \qquad [10.13]$$

and $U^* \Rightarrow Y^*$ for making bisection responses Y^* to equivalent bisection

Table 10.3

Frequency distributions of R_U for Γ V 7
with $a = 2.8, e = .10, \eta = 10, N = 100$

δU units	.25	.5	1.0	2.0
0	0	45	8	131
1	0	1	0	2
2	0	40	10	0
3	0	0	0	0
4	0	0	0	0
5	0	20	0	0
6	0	0	0	0
7	0	0	0	0
8	0	0	0	0
9	0	0	0	0

stimuli U^* ia analogous to introducing a *jnd*-like operation, on U, to generate bisection points but not direct S - R relations. The reader may protest
that if it is permissible to make constructions like [10.13], then why not
make all Y the result of $\langle U - \delta U, U + \delta U \rangle$? This idea of imprecise or
'elastic' (McGee, 1966) values is commonplace and unexceptionable in psychophysics, particularly in non-metric multidimensional scaling algorithms.
The reason is that the bisection operation requires the organism to take two
or even multiple $Y_{i:\eta}, Y_{i':\eta}$, store them, operate to construct $Y_{i,i',\eta,J}$, and
then select by modifying input U_{J+k} (some k stages later in J units of real
time) to create a new replication of $Y_{\eta,J+k}$ which *matches* $Y_{i,i':\eta}$. This must
be, in the Γ system, a higher-order operation, not something achieved by a
single passage through a recursive loop η, and while the $U \Rightarrow Y$ plots indicate what is the expected result of this, in terms of interval measurement
or some other scaling, it has to be extended by some further postulates
about the nature and stability of the storage of internal representations of
$Y_{\eta,J}, \ldots, Y_{\eta,J+k}$.

Manders (1981, p. 247) comments that there is a need for "interpretations (which) would make clear how measurement values in psychological
theory occur in psychologically significant theoretical structures". As we
have here committed ourselves to exploring a model where internal operations are set out in sufficient detail to match possible processes in the
mass action of neural aggregates, the sequential storage, comparisons and
bisections need postulated mechanisms of a related kind, and this raises
questions of how networks make comparisons between the intermediate
outputs of two subnetworks. Landahl (1945) has addressed the question,

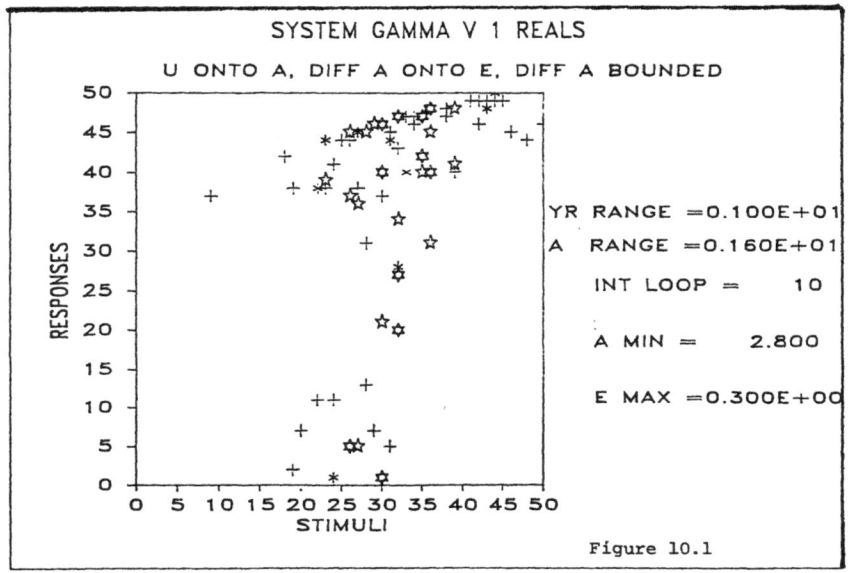

SYSTEM GAMMA V 1 REALS

U ONTO A, DIFF A ONTO E, DIFF A BOUNDED

YR RANGE =0.100E+01

A RANGE =0.160E+01

INT LOOP =   10

A MIN =   2.800

E MAX =0.300E+00

Figure 10.1

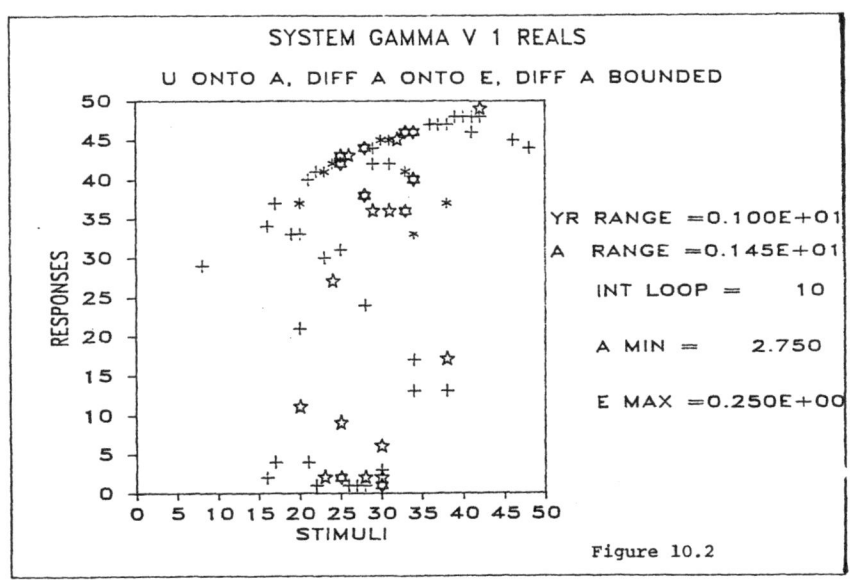

SYSTEM GAMMA V 1 REALS

U ONTO A, DIFF A ONTO E, DIFF A BOUNDED

YR RANGE =0.100E+01

A RANGE =0.145E+01

INT LOOP =   10

A MIN =   2.750

E MAX =0.250E+00

Figure 10.2

and its difficulties in the present context are touched on in Chapter 13.

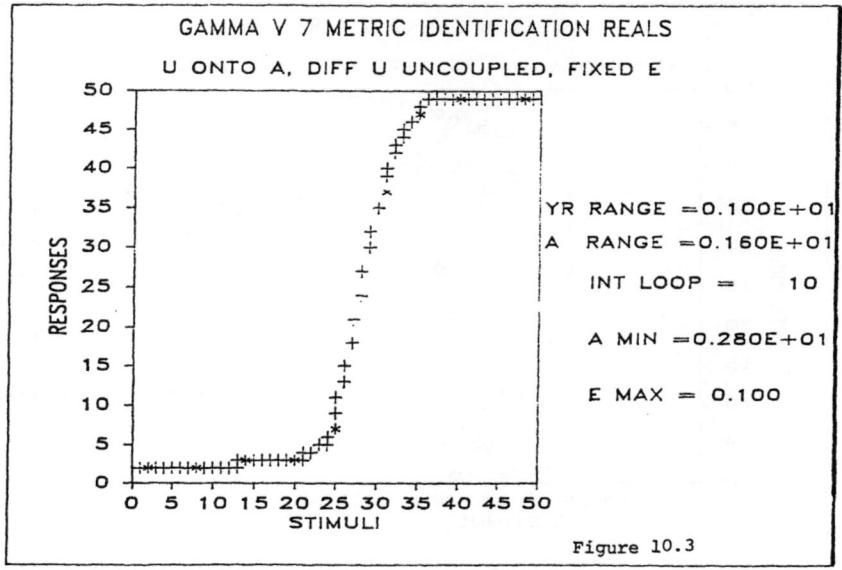

Figure 10.3

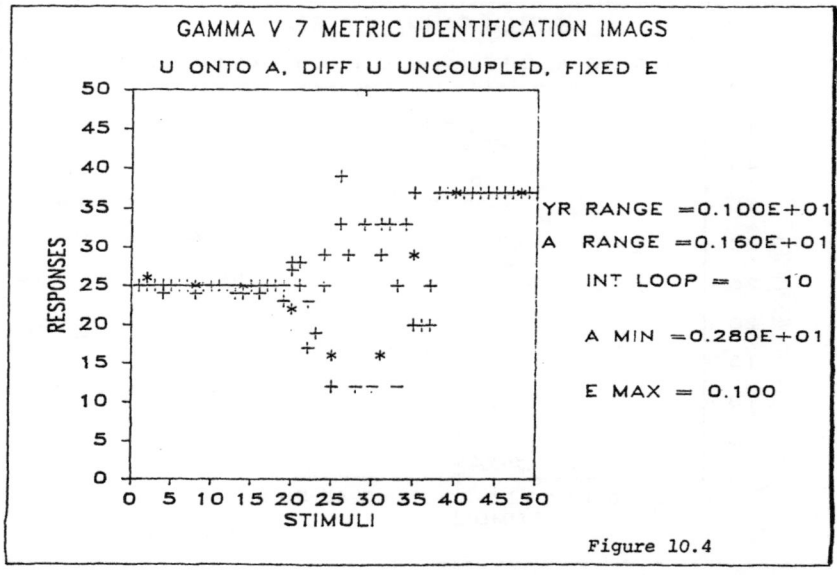

Figure 10.4

References

Beers, V. (1957) *Introduction to the Theory of Errors.* Reading, Mass: Addison - Wesley.

Borg, I. and Staufenbiel, T. (1986) The MBR Metric. *Journal of Mathematical Psychology, 30,* 81 - 84.

Eisler, H. (1980) Psychophysical similarities between rats and humans. *Bulletin of the Psychonomic Society, 16,* 125 - 127.

Fagot, R. F. A Theory of bidirectional judgments. *Perception and Psychophysics, 30,* 181 - 193.

Landahl, H.D. (1945) Neural mechanisms for the concepts of difference and similarity. *Bulletin of Mathematical Biophysics, 7,* 83 - 88.

Manders, K. L. (1981) On JND representation of semiorders. *Journal of Mathematical Psychology, 24,* 224 - 248.

McGee, V. E. (1968) The Multidimensional Analysis of 'Elastic' Distances. *British Journal of Mathematical and Statistical Psychology, 19,* 181 - 196.

Norwich, K. H. (1981) The magical number seven: making a "bit" of "sense". *Perception and Psychophysics, 29,* 409 - 422.

Pfanzagl, J. (1968) *Theory of Measurement.* Würzburg: Physica-Verlag.

Raslear, T. G. (1982) On the use of bisection procedures in animal psychophysical scaling. *Psychometrika, 47,* 95 - 99.

Schönemann, P. H. (1983) Some Theory and Results for Metrics for Bounded Response Scales. *Journal of Mathematical Psychology, 27,* 311 - 324.

Teghtsoonian, R., Teghtsoonian, M., and Karlsson, J.- G. (1981) The limits of perceived magnitude: Comparison among individuals and among perceptual continua. *Acta Psychologica, 49,* 83 - 94.

11 Analogues of SDT and Isocriterion Plots

Alternative mathematical structures for Signal Detection Theory (or for short, SDT) as a form of statistical decision theory with parametrized distributions for noise have been extensively catalogued by Egan (1975) and generally presuppose (i) two stochastic processes, N and $S+N$ on (ii) a decision axis x, (iii) responses which are either two-category or reducible to a two-category form by coalescing steps on a category scale, and (iv) two internal parameters d' and β. The parameter d' is a standardised measure of the difference between the first moments of N and $S+N$ (so that $\mu_N \neq \mu_{S+N}$ and $\sigma_N = \sigma_{S+N} = \sigma$ in x units are implicit in d'), and the second parameter β is a measure of what is variously called *bias* or *criterion* for the observer to report that a signal is present in the decision situation created by the possibilities of N and $S+N$ both being non-null on any trial.

For our current purposes we need to note that *the only observables in an SDT paradigm experiment are the two (and only two) stimulus values, U_N and U_{S+N} which are set by the experimenter, and the responses "yes" or "no"*. Note that the unobservables μ and σ are inferred and estimated parameters internal to the subject. The plots of response values which are expressed for "hits" as Prob("yes":$S+N$)/Prob($S+N$) and for "false alarms" as Prob("yes":N)/Prob(N) lie in the unit square $(0.0,1.0,0.1,1.1)$ and have been illustrated in many studies, the most recent review is by Swets (1986a,b). SDT is usually postulated as a model for relating behaviour in a detection or discrimination task to internal stochastic processes of sensitivity, and of decisions to maximise payoffs in the long run. Its importance lies in the separation of detection sensitivity (d') and response criterion or bias (β) into two independently manipulable parameters, *if the* *theory is correct* . In this present context we would like to see if the observable properties of the ROC plot in the unit square might be predictable from an SDT type experiment by using Γ without any additional assumptions beyond those already set out as [2.2] and the mappings W. It should be noted that published experimental results, whilst voluminous for ROCs, are sparse and not strongly conclusive specifically concerning the form of isocriterion curves. An isocriterion curve is one generated by fixed bias β and variable separation d' of $\mathbf{N}$ and $\mathbf{S+N}$, as distinct from an ROC (Receiver Operating Characteristic) curve which is generated by variable β for a fixed d'.

Published data are fairly consistent in their form for ROCs; examples are given by Dusoir (1975) and McCarthy and Davison (1984). Real data do not produce the clean plots of the theoretical ROCs unless there is averaging over blocks of trials and over subjects who may be regarded as homogeneous in their behaviour (Kintsch, 1968); in theory an ROC for fixed stimulus separation (d' or $U_{S+N} - U_N$) and variable bias β induced by different reinforcement schedules or feedback to the cells of the 2×2 table created by $(\mathbf{N},\mathbf{S+N}) \times$ ("yes", "no") frequency tabulation should be a monotonic function, with or without breaks or local discontinuities. A local paucity of data in the region of the diagonal in the unit square from $(0,1)$ to $(1,0)$ found by some workers has been attributed to weaknesses in experimental controls. The precise forms of the ROC and Isobias contours in the unit square have been derived from a wide range of models; it would seem that some results are more consistent with one model and others with another, but generally the assumption that $\sigma_{S+N} = \sigma_N = const.$ is false, and empirical results can and do sometimes fall below the diagonal $(0,0)$ to $(1,1)$ which in theory they should not do, because this diagonal represents a lower bound on discrimination at a chance level. Discrimination worse than chance is not predicted from SDT theory in its simplest form. However, such behaviour, rather like deliberately ineffectual behaviour, is specifically predicted by at least one model (Davison and Tustin, 1978) and by system Γ here, as will be shown.

Summarising so far, we seek to reproduce the form of ROC plots as they could actually occur in real data, and as a consequence derive related plots under fixed bias conditions, where bias has to be defined in a particular restricted sense as used below. These derived plots, analogues of ROC curves, are not smooth lines but are themselves distributions created by simulations of large samples from Γ, produced when two, and only two, close U values, mapped by W onto a_l and a_u (for a-lower and a-upper), are represented in the ROC plots in the units square. That is, *within Γ there is no special SDT model*; the limiting case arises simply because U is defined to be two-valued and mapped onto suitable high values of a.

The joint operation of W and [2.2] within Γ is to generate a range of Y_η values; e_{max} constraints induce a *one* $\mapsto$ *many* $a \mapsto Y$ relation for low e_{max}, and the nonlinearity of [2.2] induces chaos in the region of $a = 3.4$ as shown in Chapter 2. There is no simple relationship between fixed $d(a) = a_u - a_l$ and variable e in the locus of the ROC analogue produced by varying e and keeping the other parameters fixed, which would parallel that which arises in SDT by fixing d' and varying β. What we are calling an ROC analogue here is produced by varying one of the two major system parameters in [2.2], but it is not a variable bias curve in the sense of SDT.

Though Γ is fully deterministic is the present variant, Γ V 1, it generates not lines but distributions of points in the unit square of "hits" against "false alarms", and these distributions have not got moments even though one might, for descriptive purposes, in any finite data sample, fit trend lines by polynomial regression. The coefficients of such polynomials, if a is in part at least in its strange attractor mode, will not have consistent statistical estimators.

The observable scatter plots of these ROC analogue distributions vary in their degree of scatter with e_{max} in Γ V 1, and as the location of these plots within the unit square is, for fixed a_l, a_u demonstrably a function both of e and of η, the scatter varies with e and with η. As the deterministic system is not decomposable into deterministic trend and stochastic residuals, the total sample distribution is a finite realisation of the nonlinear process.

Procedure for Exploration of ROC Analogues

Simulation consists essentially of nominating *a priori* a set of constraints on a, e, η, Y_0 and then exploring over a lattice of their values within a restricted hypervolume of the system parameter space, to obtain $U \rightarrow Y_{obs}$ relations as functions of the parameters. The mapping W used here is that for Γ V 1, and

$$U_J \quad \sim U_{min} + rect(z - z_0), \quad 0 < z < 1$$
$$U_{min} \quad = 2.6 < a_{min} < 3.2$$
$$U_{max} \quad = a_{max} \quad = 3.99 \tag{11.1}$$

After U_J is generated by [11.1], it is filtered so that

$$a = \begin{cases} a_J \simeq 1 + .75a_{min} = a_l, & \text{if } U_J < (3.99 - a_{min})/2 + a_{min}; \\ a_J \simeq 1 + 1.25a_{min} = a_u & \text{if } U_J > (3.99) - a_{min})/2 + a_{min} \end{cases}$$

This gives, at this point, an input series in a which is two-valued and thus has only three possible values in $\Delta^1 a$; these are:

$$+(1 + .5a_{min}), \qquad 0, \qquad \text{or} - (1 + .5a_{min}).$$

If $\Delta^1 a_J > e_{max}$, then e is reduced to e_{max} and a_J correspondingly diminished. This induces a spread of values in a_u and a_l for some smaller values of e_{max} and of a_{min}, and $\Delta^1 a$ is consequently no longer precisely three-valued.

For a fixed $d(a)$ (corresponding to d' in SDT) which for a fixed a_{max} means $d(a)$ may be expressed as its related a_{min} value, and for a fixed η, an ROC analogue distribution is generated by varying e_{max} over the range .102 (.002) .200. Though this is not an ROC it has a topology in the same region of the unit square as an ROC and a set of such analogues can be ordered on $d(a)$ in the same way that ROCs are orderable on d', provided that they are each distinguishably different from their adjacent distributions. The plot in the unit square corresponding to the Prob("hit") against Prob("false alarm") functions is here generated by finding the range of Y_{obs} for a sample of 200 simulations for a fixed e_{max}, and then splitting Y_{obs} so that

$$
\left.
\begin{aligned}
Y_{obs} &> Y_{obs}(max - min)/2 = \text{``yes''} \\
Y_{obs} &< Y_{obs}(max - min)/2 = \text{``no''}
\end{aligned}
\right\} \qquad [11.2]
$$

and then generating the "hit" or "false alarm" probabilities as plots of a in [11.1] against "yes" and "no" in [11.2]. A "hit" is thus "yes":a_u, and a "false alarm" is "yes":a_l.

It is found that η also has a marked effect upon the location of these plots.

Features of ROC Analogue Plots

Examination of the ROC analogue plots shows some curious features of interest. See Figures 11.1 to 11.5.

For fixed η, $= 2$, 5 or 10, variable a_{min} (and hence variable $d(a)$) the distributions approach the diagonal $(0.0, 1.1)$ and with increasing a_{min} (decreasing $d(a)$) eventually cross it at some points between $(.4,.4)$ and $(1,1)$. There is a sparsity of points generated in the midrange around the diagonal $(0,1)$ to $(1,0)$ which is of interest because this is a region in which critical tests for distinguishing the predictions of low threshold theory and SDT supposedly lie (see Coombs, Dawes and Tversky, 1970); system Γ suggests that data are naturally sparse in this region so that, empirically, such critical tests would be hard to devise.

The distributions are generated for fixed a_{min} by gradually increasing e_{max}, but in a somewhat involved manner because the location of points from $(0,0)$ to $(1,1)$ is not monotonic on e in the way that an ROC is monotonic on β. Low e_{max} values generate points near $(1,1)$ and as e increases the points move towards $(0,0)$ but, as e nears 0.17 the process turns back

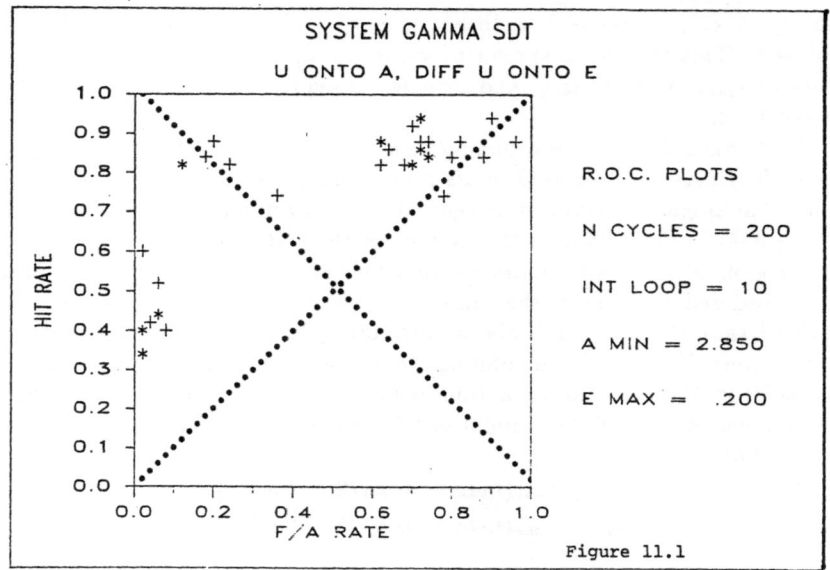

Figure 11.1

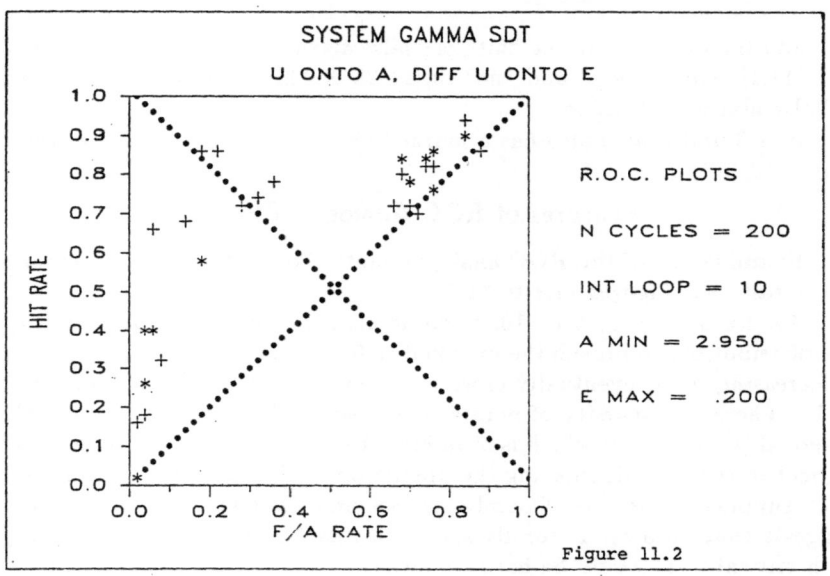

Figure 11.2

upon itself and then moves rapidly towards the region around (.5,.5) and beyond for very high a_{min}. As this is near the limit where the boundary $\{a, e\}_B$ on acceptable values of ae in [2.2] are exceeded such points may not correspond to any stable biological process implied in detection tasks.

If, in real experiments, instructions to bias responses operate in some way on the modifiable parameters on Γ, then such instructions either have an intricate effect upon e, perhaps asymmetrically with respect to direction of change of a_J to a_{J+1}, or e_{max} is bounded at around 0.17. One could advance biological arguments that a system with such bounds is self-protecting against overload, but a re-examination of real data is probably necessitated by the question raised here, because inducing, by instructions to subjects, biased responses which should by SDT theory be near to $(0,0)$ is not generally done. Hence data which would be critical for this question may not exist or may have been averaged out in published results and therefore lost.

The effect of increasing η is complicated for low values of η but beyond $\eta = 10$ is negligible. If η is low, for $Y_0(\text{Re})$ set at the midpoint of a_l, a_u the dense part of the ROC analogue moves from $(0,0)$ towards $(1,1)$; that is to say, contrariwise to the effect of increasing e_{max}. This suggests strongly that if η were coupled to e as a simple linear relation or a scalar multiplier, then a plot over the whole unit square above the diagonal $(0,0)$ to $(1,1)$ could be generated in a simulation with the degrees of freedom of Γ actually reduced by one. There is thus a topological similarity, in the plots within the unit square generated by Γ V 1 with a bimodal U input, to the predictions of SDT, in the sense that the input parameter $d(a)$ affects the general location of the distribution, and the two parameters which can be varied independently of a, namely e and η, generate the points at different loci in a distribution which appears potentially continuous but not uniformly dense.

Analogues of Isobias Curves

In a family of distributions from Γ in the unit square, for fixed $d(a)$ and variable e_{max} and η, one might construct the corresponding families of plots for the cases, say, of fixed η, and variable e_{max} and variable $d(a)$. This would give a loose analogy to the process of obtaining isocriterion curves from ROC curves in SDT, but the present system parameter space is more complicated and does not transform in the same way to produce a family of curves corresponding in the way that an isobias family has one and only one corresponding ROC family.

Instead we may resort to at least one of the definitions of bias used in SDT and attempt to derive isobias plots from variable a_{min} in Γ V 1 directly. It must not be expected that this will necessarily yield a to-

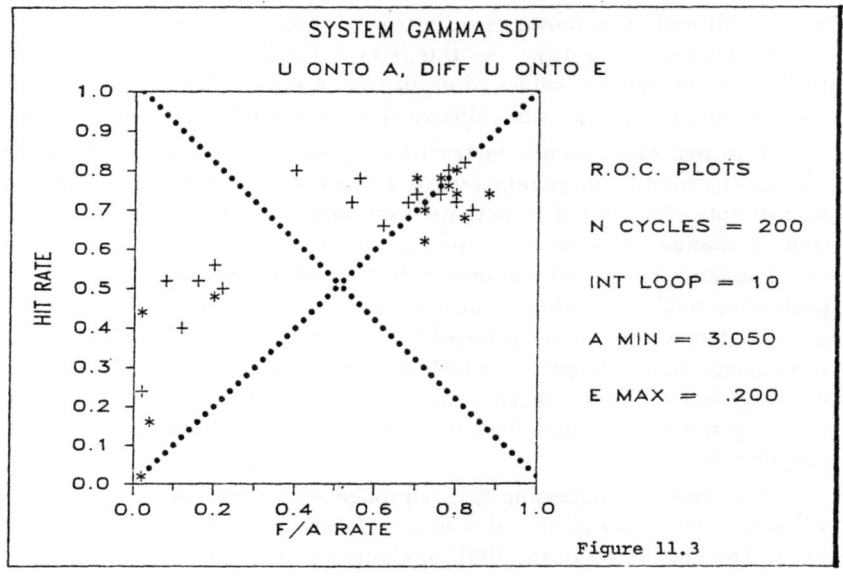

Figure 11.3

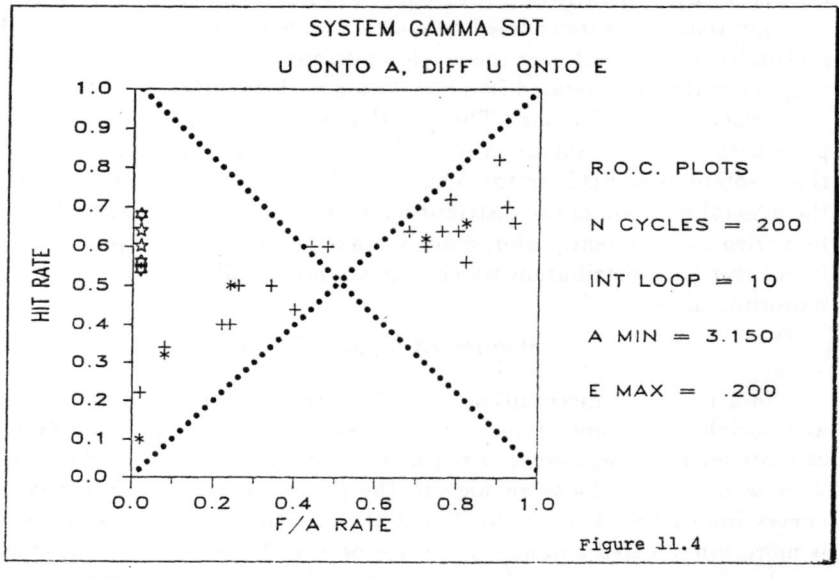

Figure 11.4

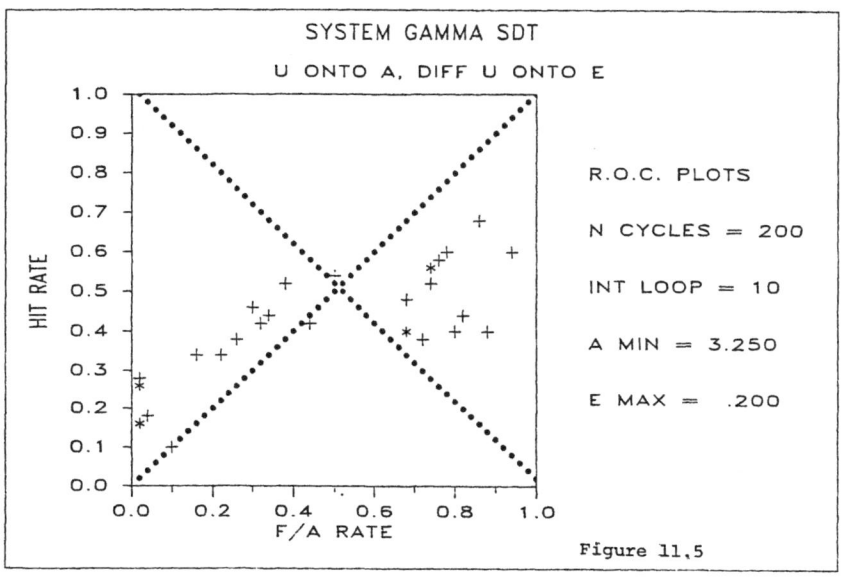

Figure 11.5

pography of isobias analogue distributions which resemble reported data (although such are rare and inconsistent) because no external feedback to induce bias has any representation in Γ, but it will eventuate that some correspondences do emerge.

There would appear to be at least seven different definitions of bias in the literature of SDT as reviewed by Dusoir (1975, 1980, 1983) and none of them fits *all* reported data; some on closer examination are untestable or fit little if anything. Indeed the diversity of shapes and locations of apparently lawful isobias plots (*lawful* meaning here amenable to a low-order polynomial representation on an adequate data sample), from subject to subject, for the same experimental conditions, is sufficient to invalidate any model based on the stochastic distribution and decision theory approach used, for example, by Egan (1975) and later special cases (Craig, 1976, Kornbrot, 1984, Treisman and Faulkner, 1984).

There is no point in reviewing any of these models because (a) they do not generally fit data and so there is no purpose in seeking to replicate their predictions from the nonlinear bases of this study, and (b) our objective here is to simulate real data properties and not those of other hypothetical models fitted to mean pooled data. It is therefore expedient to have a basis for generating a diversity of isobias-type plots, notwithstanding that we have already shown that the ROC analogues are generated by parameter changes in a_{min}, and in e_{max} which is not by definition a bias parameter. In other words, if bias as an external conditioning variable does not operate to move subjects up and down an ROC-analogue zone in the unit square, then it must, in this model, do it not by there being an internal homunculus operating its biases in some quasi-rational reflection of normative statistical decision theory, but instead by the instructions or training we apply to a subject in the guise of bias-inducing instructions or reinforcement schedules operating on the only places that stimuli can modify the internal dynamics of the system, namely upon a, e, η or weights $W(x)$ on the mappings of U and $\Delta^1 U$ onto a and e.

Features of Isobias Analogue Plots

The SDT analogue of partitioning the (a, Y_{obs}) plot we have used gives as a "bias" R_b which for a given sample of observations has the ratio

$$\frac{(\mathbf{n}(\text{falsealarms}) + \mathbf{n}(\text{truenegatives}))}{(\mathbf{n}(\text{falsenegatives}))}$$

or

$$R_b = \mathbf{n}(a_l)/\mathbf{n}(a_u) \mid Y_0 \qquad [11.3]$$

This ratio ranges from sample to sample and is not uniformly 1, due to the action of the filter W.

It is required to find the coodinates in the unit square of Hits × False Alarms that will maximise the probability of a correct decision, given R_b, on the assumption that the observer is using an appropriate maximising strategy (Egan, 1975, pp 20 - 22, 44). As the data are not sufficiently dense to use precise R_b values in generating isobias plots, we have used $R_b \pm \delta R_b$ as a segment of the R_b continuum and collected all coordinates that satisfy the segment limits onto one plot. Again, note that the isobias plots are now distributions and not lines as in the SDT paradigms. For simulation we have used

$$R_s = log_{10}(10R_b) \qquad [11.4]$$

so that $R_s = 1$ for the no-bias condition where $R_b = 1$, but a wider range of bias values can be explored for clarity.

Induction of bias by extraneous reward and not by stimulus properties is more usual; indeed the induction of bias in infrahuman species appears to be effectively achieved only by reinforcement schedules (McCarthy and Davison, 1984). Hence the R_s plots produced here, as in Figures 11.6 to 11.10, whilst a system characterisic of interest, are not necessarily predictions of the empirical consequences of selective reinforcement unless such

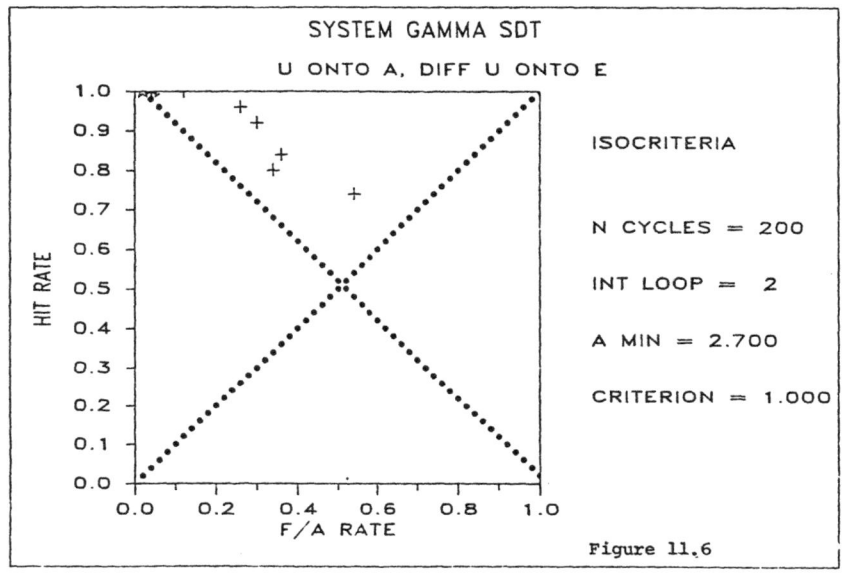

Figure 11.6

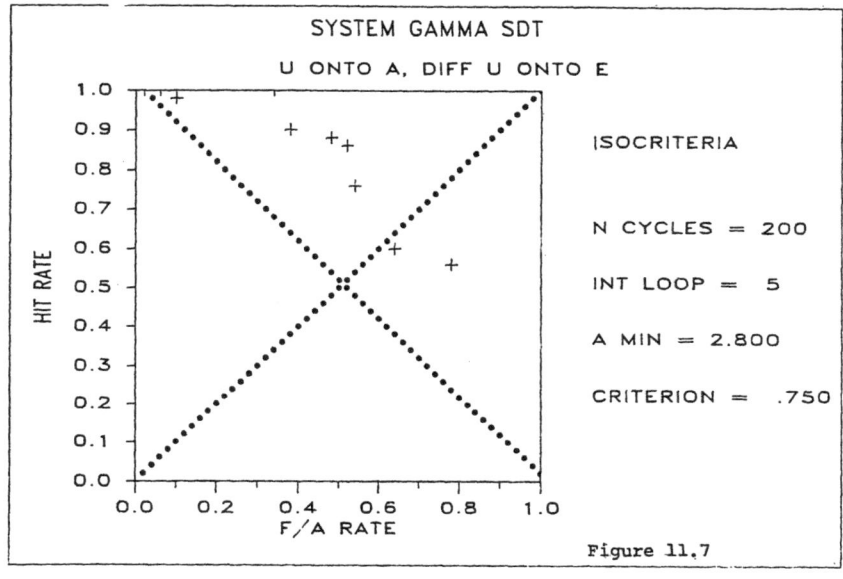

Figure 11.7

reinforcement is in. and only in, our sense of bias [11.3], like a selective filtering of input prior to W in Γ. As we have departed from the parameter space of the SDT model and do not know if any regular homomorphism between the SDT(d', β) and the $\Gamma(a, e, \eta, R_b)$ spaces could exist, the question is open. What is of interest is that the R_s plots do have some resemblance to the data of McCarthy and Davison (1984) in the way that they can cross over the (0,0, 1,1) diagonal and then turn backwards. A crossing over like this does appear also in some data illustrated by Dusoir. This is not predicted from the gaussian stochastic SDT model[1].

Behaviour of the Corresponding $Y_{obs}(\text{Im})$

The variable used here is $Y_{obs}(\text{Re})$; it is of interest as a check to see what happens to the other part, $Y_{obs}(\text{Im})$ if this is used instead of the Real part as the input to that part of the simulation creating the Hit $\times$ False Alarm plots. The Imag values have, as earlier, to be normalized over a comparable range and multiplied by a factor of about 10^8 in order to be comparable.

Using the same set of a, e, η parameter values which produced the plots resembling ROCs for various a and e combinations, one immediately obtains plots of Hit $\times$ False Alarms based on the Imag component, but these are now all very similar to one another, and cluster near the diagonal from (0,0) to (1,1), like random behaviour. In SDT theory these distributions now correspond to guessing behaviour or to null information transmission; there is a strong tendency to cluster at the ends (0,0) and (1,1) and the scatter about the diagonal increases with η , but there is apparently also some interaction between a and η in this regard. This is shown in Figures 11.11 to 11.13.

The corresponding isocriterion analogue plots, in Figures 11.14 to 11.16, lie, again with varying scatter, along the other diagonal (0,1) to (1,0) and again tend to cluster at the corner points. This suggests that the Imag component is insensitive to the value of the criterion R_b used, though

---

[1] A referee asked, why not consider the possibility that responses below the diagonal are due to a 'perverse' subject who swaps over the response labels 'signal' and 'noise' after detecting something. This suggestion is rejected for two reasons; (i) it is an added postulated mechanism, which is unnecessary in Γ, though it is needed to salvage SDT, and we are not in the salvage business, (ii) if one wished to add on this perverse response selection process then additionally one would have to show precisely why it occurs with detectable consequences only in some parts of the square plots, and then express this mathematically. Once started on this slippery path one ends up with special ancillary hypotheses for any point in real data which seems curious or anomalous.

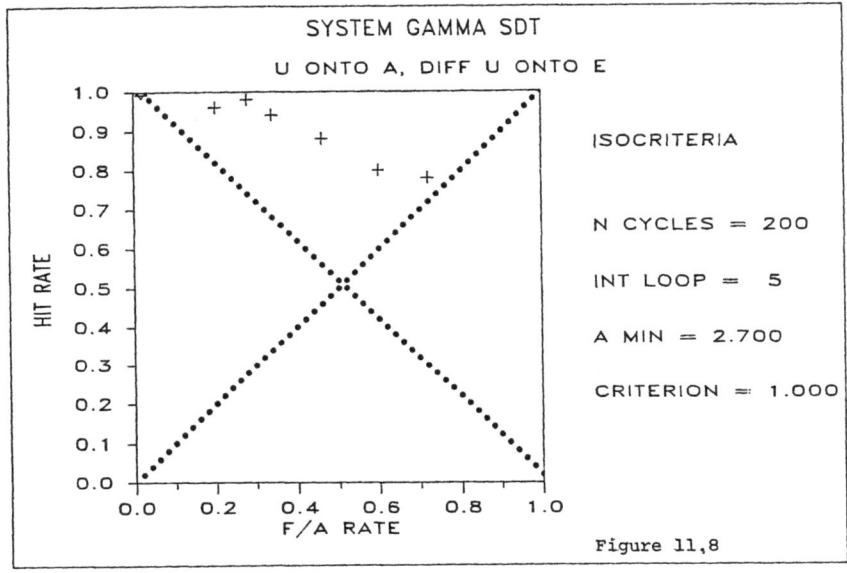

Figure 11,8

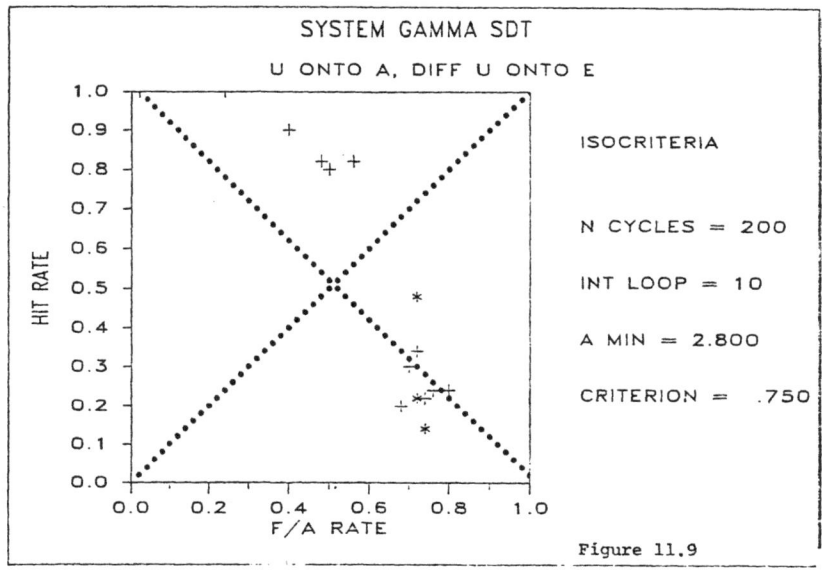

Figure 11.9

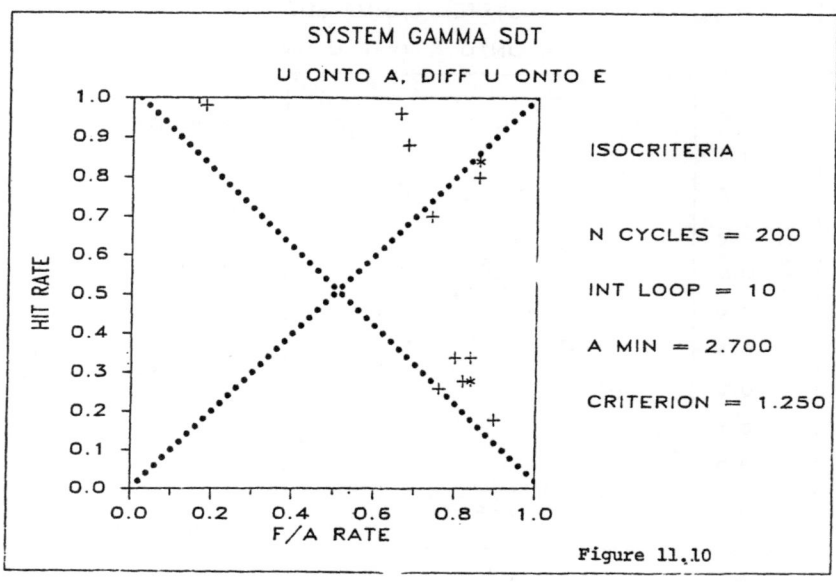

Figure 11.10

obviously in the light of previous comments about the number of possible alternative ways in which criterion bias might be defined, this finding can only relate to the interpretation used here.

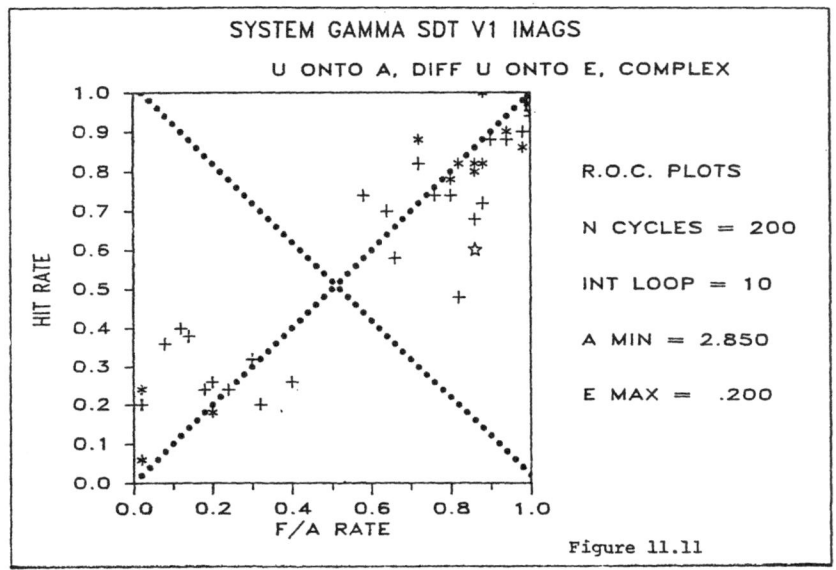

Figure 11.11

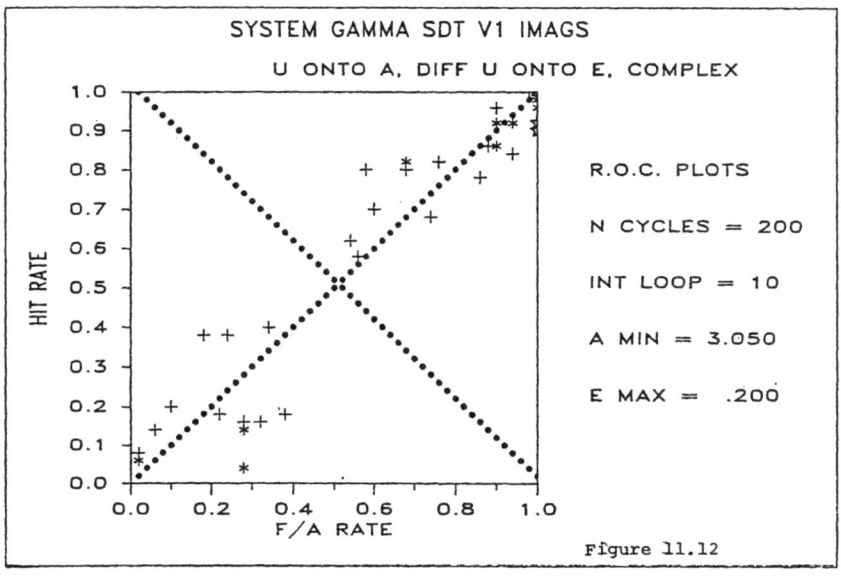

Figure 11.12

References

Coombs, C. H., Dawes, R. M. and Tversky, A. (1970) *Mathematical Psychology: An Elementary Introduction.* New Jersey: Prentice Hall.

Craig, A. (1976) Signal recognition and the Probability-matching Rule. *Perception and Psychophysics, 20* 157 - 162.

Davison, M. C. and Tustin, R. D. (1978) The Relation between the generalized Matching Law and Signal Detection Theory. *Journal of the Experimental Analysis of Behavior, 29* 331- 338.

Dusoir, A. E. (1975) Treatment of Bias in Detection and Recognition Models: a Review. *Perception and Psychophysics, 17,* 167 - 178.

Dusoir, A. E. (1980) Some Evidence on additive Learning Models. *Perception and Psychophysics, 27,* 163 - 175.

Dusoir, A. E. (1983) Isobias Curves in some Detection Tasks. *Perception and Psychophysics, 33,* 403 - 412.

Egan, J. P. (1975) *Signal Detection Theory and ROC Analysis.* New York: Academic Press.

Guevara, M. R., Glass, L., Mackey, M. C. and Shrier, A. Chaos in Neurobilogy, *IEEE Transactions on Systems, Man and Cybernetics, SMC-13,* 790 - 798.

Kintsch, W. (1968) The Experimental Analysis of Single Stimulus Tests and Multiple Choice Tests of Recognition Memory. *Journal of Experimental Psychology, 76,* 1 - 16.

Kornbrot, D. E. (1984) Mechanisms for Categorization; Decision Criteria and the Form of the Psychophysical Function.*British Journal of Mathematical and Statistical Psychology, 37,* 164 - 198.

McCarthy, D. and Davison, M. (1984) Isobias and Alloibias in Animal Psychophysics, *Journal of Experimental Psychology: Animal Behavior Processes, 10,* 390 - 409.

Swets, J. A. (1986a) Indices of Discrimination or Diagnostic Accuracy: Their ROCs and Implied Models. *Psychological Bulletin, 99,* 100 - 117.

Swets, J. A. (1968b) Form of Empirical ROCs in Discrimination and Diagnoistic Tasks: Implications for Theory and Measurement of Performance. *Psychological Bulletin, 99,* 181 - 198.

Treisman, M. and Faulkner, A. (1984) The Effect of Signal Probability and the Shape of the Receiver Operating Characteristic given by the Rating Procedure. *British Journal of Mathematical and Statistical Psychology, 37,* 199 - 215.

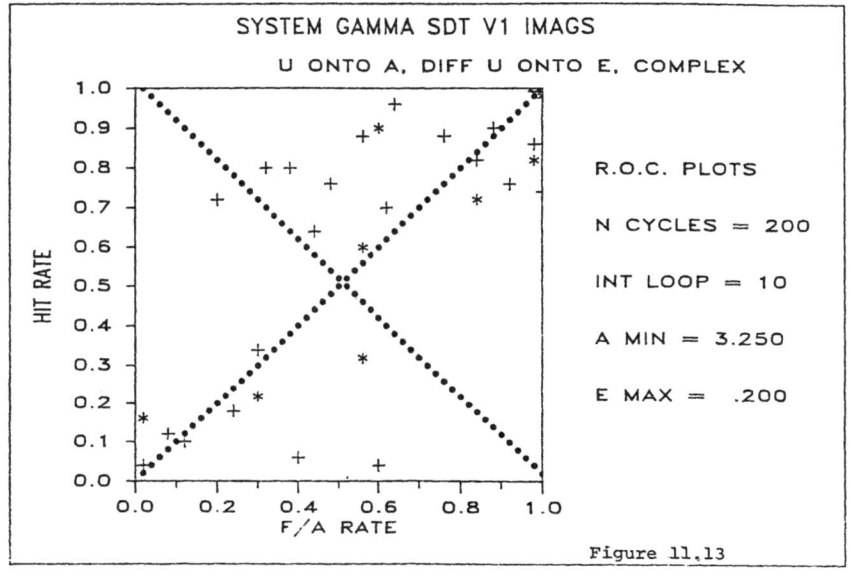

Figure 11.13

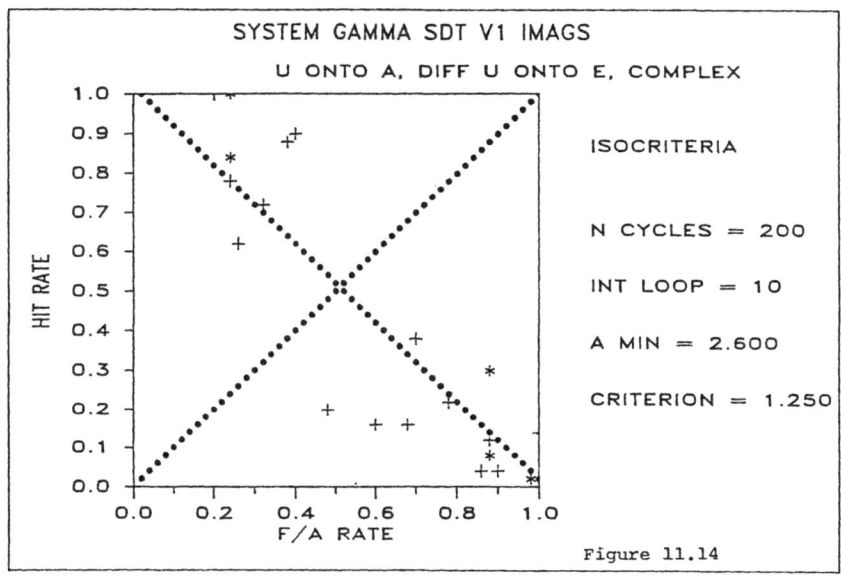

Figure 11.14

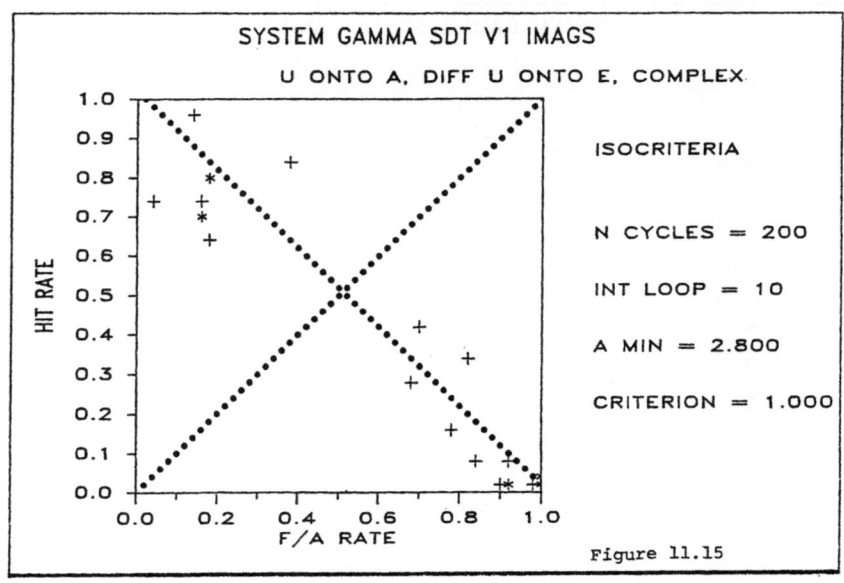

SYSTEM GAMMA SDT V1 IMAGS

U ONTO A, DIFF U ONTO E, COMPLEX

ISOCRITERIA

N CYCLES = 200

INT LOOP = 10

A MIN = 2.800

CRITERION = 1.000

Figure 11.15

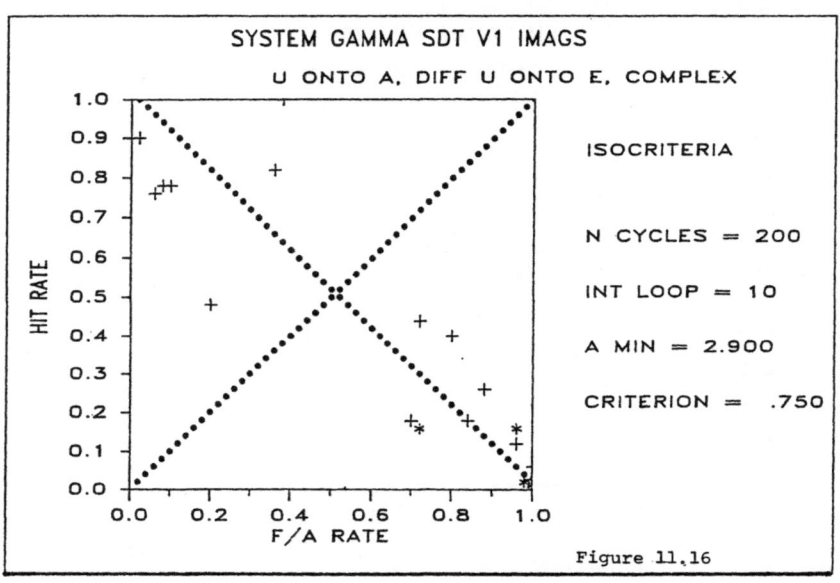

SYSTEM GAMMA SDT V1 IMAGS

U ONTO A, DIFF U ONTO E, COMPLEX

ISOCRITERIA

N CYCLES = 200

INT LOOP = 10

A MIN = 2.900

CRITERION = .750

Figure 11.16

12 Range and Transposition Effects

It has been established that the simplest psychophysical models are not generalizable without regard to contextual conditions. For example, the exponent β in [3.2] is not properly taken as a constant, but is manipulable even in situations where the equation approximates to data (Ahlström and Baird, 1987).

Transposition

We may consider two main ways in Γ in which sensitivity to stimulus intensities can be varied:

(a) The mapping of a range of U values onto some closed segment of the a continuum can be changed.

If the organism is hypersensitive, then a weak environmental stimulus is highly destabilizing, so that a low U range maps into a high a range. This, because of the nonlinear dynamics of [2.2], for fixed e and fixed η, implies a steep CNO and a difference threshold which is proportional to e.

If a high U maps onto a low a range, the situation is hyposensitive; weak stimuli have but a slight perturbing effect, yet the ogive tends to be very flat in the low a region so that the difference threshold is relatively large.

The Crozier quotient stability phenomenon (Crozier and Holway, 1950) which only holds for some monotone transformation of the stimulus continuum U, implies that the value in physical units of the absolute threshold is proportionate to that of the difference threshold, as though one single central sensitivity process mediated both the processes of discrimination and detection. If this is so, then high U mapping onto low a implies a high absolute threshold. In turn this implies a high difference threshold, which is a small slope of the psychometric input-output, $U \mapsto Y_{obs}$, function in the vicinity of the a value corresponding to the absolute threshold in U units. Contrariwise, a low U mapped onto a high a implies a low absolute threshold, and this is consistent with the existence of a low difference threshold.

Variation in the slope of the input-output function, with a fixed point of entry onto the a continuum, may actually be achieved by varying e for a fixed η, as will be shown in the next section. Figures 12.1a and 12.1b show possible forms.

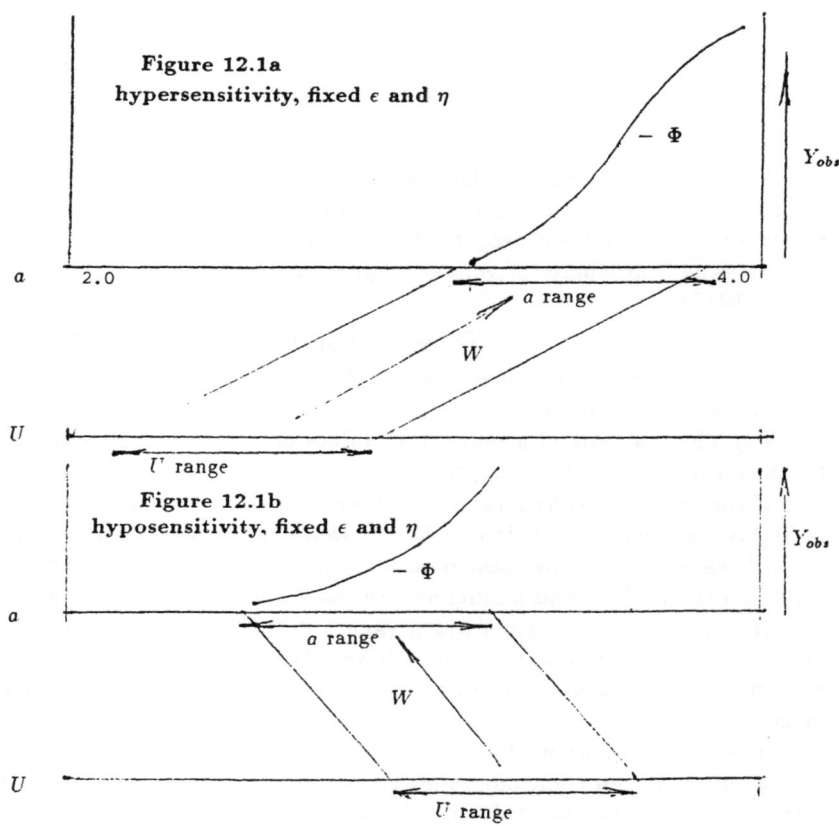

Figure 12.1a
hypersensitivity, fixed ϵ and η

Figure 12.1b
hyposensitivity, fixed ϵ and η

(b) The mapping of u onto a is fixed and lies over a wide range of a, and the changes in the sensitivity of the organism are reflected only in the parameter e. Sensitivity in this case may be thought of as almost inversely proportional to internally generated system noise which is shown in $Y_j(\text{Im})$, optimal sensitivity is associated with optimal levels of internal chaotic dynamics. It does not follow, however, that progressively greater e implies lower and lower sensitivity; the reasons for asserting this are expanded upon later.

Using Γ V 7, only within the range $.10 < e < .18$ does the CNO form of the $U \mapsto Y_{obs}(\text{Re})$ plot arise in a clearly recognizable shape; so in this region we might fit by the method of moments a gaussian ogive to the input-output function, and the regression of one upon the other takes up not less that 99.2% of the linear regression variance. However, the slope of this regression of the system gain function on a corresponding CNO is $1.13 < b < 1.80$, so a transformation of the a (or U) axis would be needed to bring the form into a CNO and not something like a logCNO. Over the e range cited the difference threshold increases about 2.5 times, and the a associated with the mean Y (L_{50} for Y_{obs}; a) drops from $3.52a$ to $3.24a$. Simultaneously the Y_{obs} response range increases from $.530$ to $.615$ as the difference threshold in a units increases; e affects the two measures of discriminative sensitivity in a different manner; they are in fact negatively coupled over the range of e, if mapping is onto the total a range from U. This situation arises because the family of input-output curves has a common top limit and runs up to it at different rates; increasing e increases responses to lower U values for a fixed U onto a mapping. It may be deduced that varying e for a fixed $U \mapsto a$ is *not* the basis of Crozier quotient invariance. See Figures 12.2 and 12.3.

(c) Locking U onto approximately $a_{min} = 3.0$, with fixed η, for a narrower range of U to cover only the ogival steepest part, will, as e is increased, give the changes in shape which resemble results by Sarris *et al* (1985). See Figures 12.4 to 12.7.

This range of changes in the shape of the input-output function is actually a narrower vertical slice through the pattern in Figure 12.3. It follows that the Frame of Reference (F of R) experiments are incompatible superficially with the Crozier invariance property. There is, however, a basic distinction in that the Crozier result holds as near-threshold sensitivity itself varies between observers, or between conditions for a given observer, whereas the F of R effects are markedly suprathreshold and imply nothing about the detectability of differences; instead they are expressed in terms of the responses chosen by the observer to represent a given stimulus within a variable set of contextual stimuli. In this situation, where a task is to respond relatively to members of a closed set of stimuli, every stimulus is part of the context of every other.

One may thus postulate that the Crozier quotient invariance effect

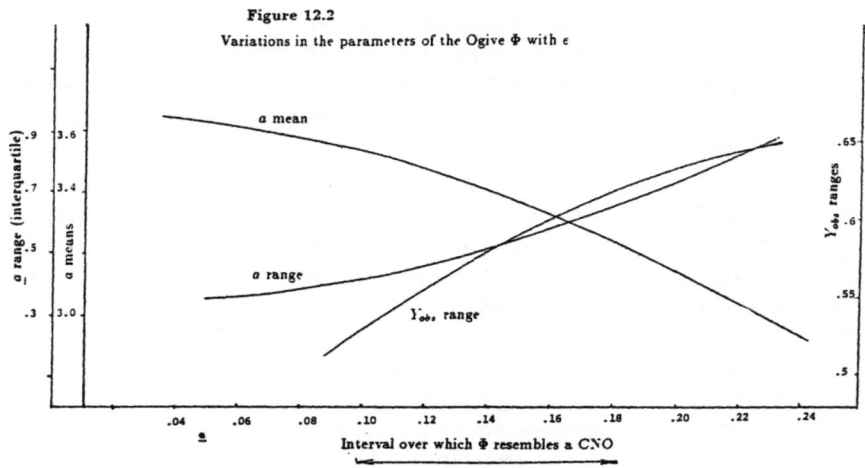

Figure 12.2

Variations in the parameters of the Ogive Φ with ε

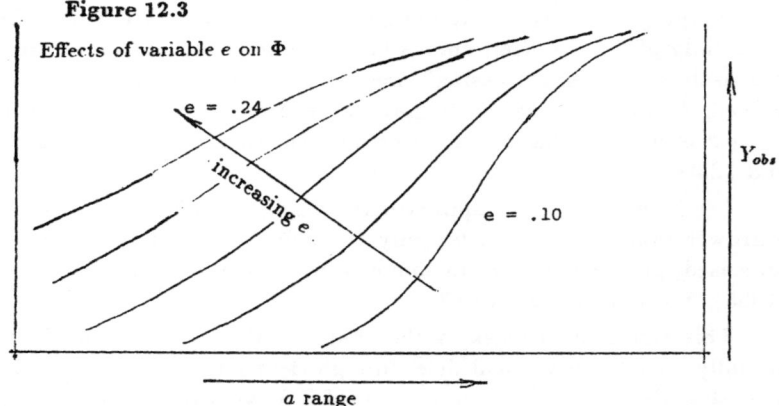

Figure 12.3

Effects of variable e on Φ

is not in fact generated in a manner which is analogous to the F of R effects. Whilst the Crozier invariance property implies that the detection and difference thresholds are coupled, the F of R results say nothing about thresholds, but simply predict that a shift in the stimulus range temporarily induces the observer to operate in a different part of the $U \mapsto a \mapsto Y_{obs}(\text{Re})$ system. It should be noted that F of R results seem to be inconsistent from one experiment to another (Wilson, Mackintosh and Boakes, 1985). In some cases a shift in the stimulus range induces transposition, in other cases if the shift is greater then the responses become absolute; individual stimuli retain the same category labels on a scale of relative magnitude. Some species do not show transposition at all, apparently. This diversity of behaviour might be accomodated for by postulating that a small shift in the range of U induces a compensatory shift in e within the observer which holds the CNO-like form about the same. A large shift in U is too much for the internal dynamics to compensate against, and a consequent shift in the form of the input-output plot is induced. The problem with this sort of explanation is to define, for a given observer, what is a small shift and what is a large shift in U.

So, varying e would produce an analogue of the F of R results, whereas varying U onto a would produce, for fixed e, the Crozier invariances. That is, threshold and input-output transform changes need to be represented within Γ in quite different regions of the system parameter space.

It is perhaps counter-intuitive that F of R could operate through shifts in e, if we were to think of e as simply a measure of central noise in the system, as in Chapters 4 and 11 for example. In $Y_{obs}(\text{Im})$ distributions corresponding to the $Y_{obs}(\text{Re})$ input-output plots which resemble CNO considered here, there are three discernable phases; for low a the Im component is fairly constant, then variability in $Y_{obs}(\text{Im})$ with increasing a which suggests a chaotic phase is associated with rapidly increasing $Y_{obs}(\text{Re})$, and then this in turn is followed by cycles and bifurcations (see Figure 12.4). This change in phases is not within the system loop governed by [2.2], but in the outer relationship of Y_{obs} for fixed η as a is increased by small increments. Decreasing or increasing suddenly the stimulus input will change the relation of Y_{obs} to a; the most chaotic region is associated more probably with the $Y_{obs}(\text{Re})$ on a mapping which resembles a CNO. Frame of Reference displacement can thus be thought of as a displacement from the system's maximally chaotic internal state, when the chaos is induced by second-order changes in a. To see if this idea is plausible we have to move to more general ideas about the stability of chaotic systems. Some properties discussed by Prigogine and Stengers (1984) are helpful.

There are thus two main problems:

(i) Should stabilization occur as an intrinsic property of nonlinear

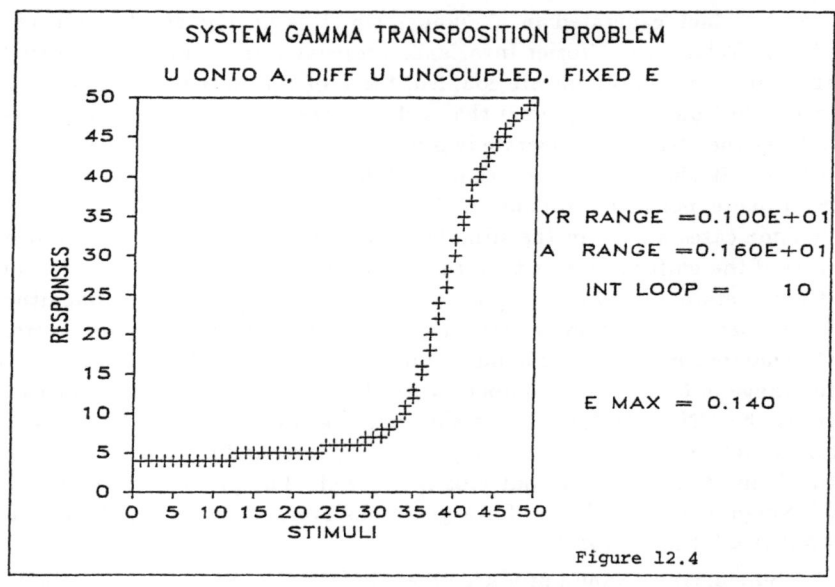

Figure 12.4

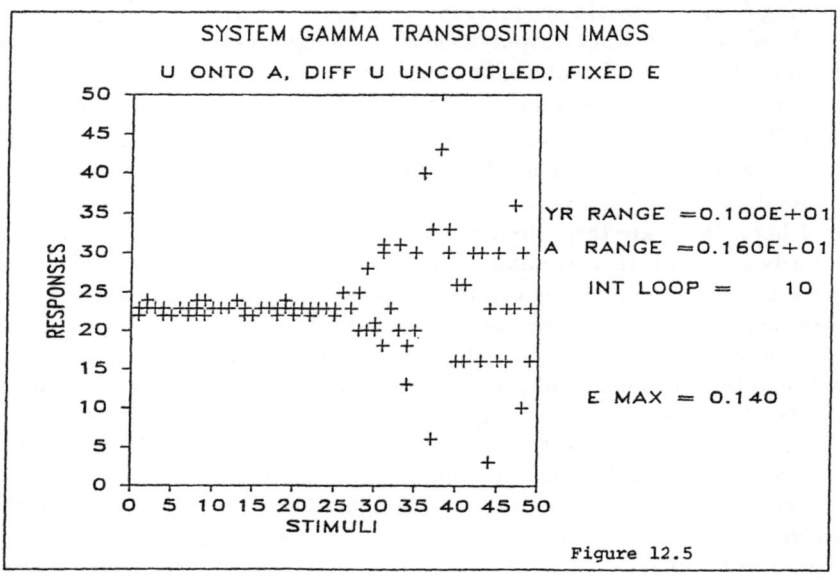

Figure 12.5

systems that can move into and out of chaos, without therefore the need to invent some additional but unobservable stabilization mechanisms, as Treisman (1985) attempts ?

(ii) If the answer to (i) is yes, are there quantitative properties deducible from the Γ process about the possible upper and lower limits of restabilization rates ?

Commenting on (i): psychologically it has been suggested by various workers such as Helson (1947) and later Parducci (1984) that the behavioural $S - \mathcal{R}$ mappings tend to a centralised sort of stability, with the relative range and frequency of the elements of U being crucial determinants in forming the $U \rightarrow a \mapsto Y_{obs}$ mapping[2] In some of these earlier models Y has its observable response output in the form of category scale usage only.

In system theory change is towards an 'attractor state' (Prigogine and Stenger, 1985, p. 124) which has maximal probability amongst the states of the system. Such a state has macroscopic disorder and symmetry, and corresponds to the majority of statistically possible but unobservable microscopic states within the system. Once the system is in an attractor state it will only fluctuate around it; effectively it is trapped in the dynamics of the neighbourhood of the attractor state. The suggests that the defining characteristics of an attractor state may be identifiable, as showing maximum output entropy. As against this, theorizing about what happens when the system is displaced from its attractor state and thus from its associated intrinsic stability is equivocal, because it is not clear if the states into which the $S - \mathcal{R}$ system is displaced by F of R changes are to be regarded as near-to-equilibrium or far-from-equilibrium.

A Case Study in Range Effects

The problems of the psychophysics of mixtures of two sometimes discriminable stimulus dimensions are particularly interesting in olfaction, partly because relatively little is known and partly because olfactory stimulation has strong autoregressive consequences in the observable responses. The disentangling of the sequential dynamics of odour mixture perception (Gregson, 1982, 1983a, 1984a, 1986, Gregson and Gates, 1985) might thus serve as a testing ground for the development of a multidimensional psychophysics which could, with appropriate modification, be extended to the study of other senses where the time course of events is considerably faster.

Evidence cited so far suggests that (a) aqualitative intensity and qualitative intensity are mediated by different transfer functions, and (b) the information transmitted in a stimulus-concentration – response-intensity

---

[2] Though obviously these earlier theories do not have an intervening variable like a, but use some construct like the geometric mean of U.

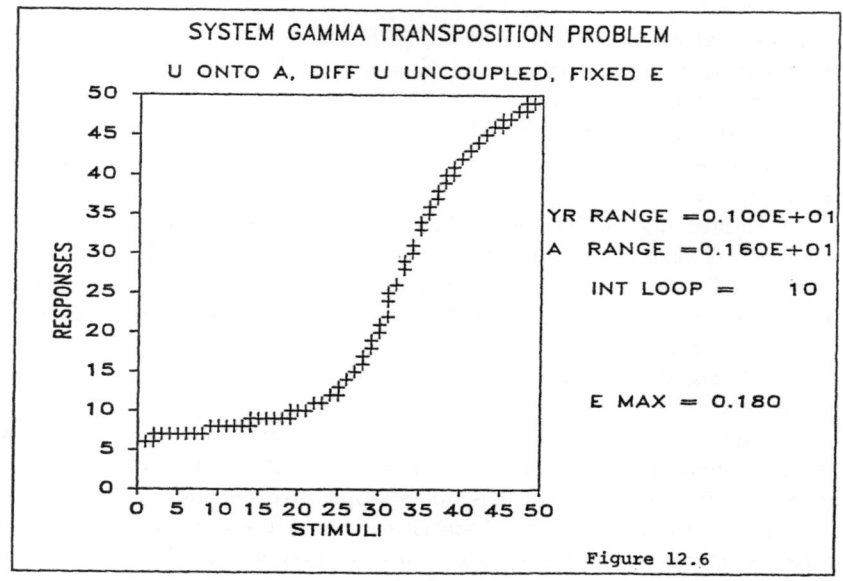

Figure 12.6

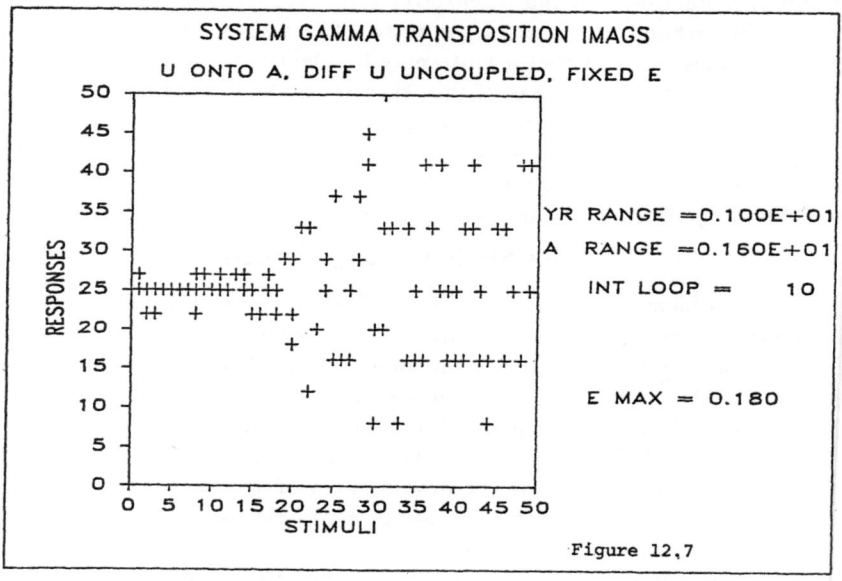

Figure 12.7

task is associated with the modulus of the largest dynamic root of the transfer function, (c) in low intensity stimulation the major dynamic root is conjugate complex, implying the presence of a convergent oscillating process in time, (d) when there are two such conjugate roots in 2-input 2-output experiments they are not necessarily associated $1 \Leftrightarrow 1$ with the stimulus dimensions, and (e) for higher intensities, or for a wide range of intensities in the same experiment, a negative real root will also be present. Transfer function analysis, z transform representation, and the use of root locus diagrams for dynamic component identification are treated in Bennett (1979), Gregson (1983a), Robinson (1983), and numerous engineering texts on control theory.

These findings from dynamic analysis obviously do not employ the usual psychophysical methods which stop at the zero-order[3] statistics of the $S - R$ average relations. The simplest attempts at summarising the effects of one sensory input dimension upon another either use the mean level of a second dimension $\bar{u}_2$, as a modifier of the slope m_1 of the regression of the first dimension's mapping of y_1 on $u_1 \mid \bar{u}_2$ (Cain and Drexler 1974, Cain 1975), or another approach which goes back over 50 years, which is to note when a strong component u_1 completely obscures a weak component u_2 so that $y_2 = 0$. Such data may be summarised (Lehky, 1983) in a plot of equal intensity lines $y_1 + y_2 = c$ on the u_1, u_2 axes as such curves have a peculiar form with marked end effects, and depart significantly from a straight line $u_1 + u_2 = c'$. These approaches are essentially static, resting on averaging responses over a series of stimuli, and are completely confounded with sequential interactions, both intra- and interdimensionally.

In the example used here we were influenced by two considerations, (i) to see if the dynamic analysis of the empirical output of 2-input olfactory psychophysics can be systematically related to target component intensity and a second weaker (background) masking or noise component, and (ii) to compare the factorization of the theoretical impulse response coefficient spectrum from the Γ V 7 case with those from various conditions in the real experimental data. It is a necessary but not sufficient requirement of a nonlinear psychophysical model that it generates from a random U input series an output series Y with the autoregressive properties associated with real empirical data. It is worth noting, in this context, that Γ V 7 is only for one-dimensional input, and there are thus limits on how closely the analogy to the 2-input psychophysics experiment may validly be pursued. We return to this question in Chapter 13.

---

[3] Zero-order here means something calculated without regard to the time sequence of observations involved. Such an approach is effectively a filtering-out of any dynamic information.

Generic Model

To obtain the transfer functions for u_j (stimulus values) to y_j (response values) over all the 22 individual data sets in a fully comparable form we set up a generic model which has sufficient autoregressive and noise structure to encompass the diversity of results. The alternative is to find interactively a minimum-order model for each subject's data separately and then take the union of all the coefficient sets as an envelope model for the four groups, which is more laborious; here there is sufficient analogous dynamic structure in the data, revealed by a preliminary sampling, to hazard using one model in which for a given subject some but not all coefficients will be negligible.

The form of a transfer function, in z operator[4] notation, where j is an integer denoting the trial number, is (Young, 1984, eqn. 6.29)

$$y_j = \frac{B(z^{-1})}{A(z^{-1})} \cdot u_j + \frac{D(z^{-1})}{C(z^{-1})} \cdot e_j \qquad [12.1]$$

where

$$x_j = \frac{B(z^{-1})}{A(z^{-1})} \cdot u_j \qquad [12.2]$$

is the system model,

$$\varepsilon_j = \frac{D(z^{-1})}{C(z^{-1})} \cdot e_j \qquad [12.3]$$

is the noise model, and hence

$$y_j = x_j + \varepsilon_j. \qquad [12.4]$$

To proceed we have assumed all the roots of $C(z^{-1})$ lie outside the unit circle (a condition which all the data sets used here satisfied) and

$$A(z^{-1}) = 1 - a_1 z^{-1} \qquad [12.5]$$

$$B(z^{-1}) = b_0 + b_1 z^{-1} + b_2 z^{-2} + b_3 z^{-3} + b_4 z^{-4} + b_9 z^{-9} \qquad [12.6]$$

where b_0 represents the instantaneous operation of u on y. It is to be expected, and can easily be shown, that b_0 correlates highly with $exp(m)$ where m is the slope of the zero-order regression of $logR/logS$. They

---

[4] The operator z is a one-step shift in time; it corresponds to the backward shift operator **B** used by Box and Jenkins (1970) which is met in some computer packages for time series analysis.

are not, however, the same thing, because their definition and estimation procedures are fundamentally different.

Experiment

Method

A continuous-flow dilution olfactometer (for design details see Gregson, 1984c) was used to produce a series of 56 two-component odour mixtures. The concentrations of Bergamot (Bergamote 61, a Givaudan nontoxic synthetic analog of the natural oil) and n-Butanol in these mixtures were varied randomly over a rectangular distribution, the whole series being generated by computer and analysed on-line for zero-order statistics. Four different conditions were used, to cover two intensity ranges of Bergamot as the target odourant, and two levels of n-Butanol as the background masking odourant.

Conditions

100/1	Butanol at 100 ml/min,(equivalent to 0.1 in 150 l/min or 6.6×10^{-4}sat.vap/air), Bergamot range 50 – 550 mls/min.
100/2	Butanol at 100 mls/min, Bergamot range 560 – 990 mls/min,
300/1	Butanol at 300 mls/min, Bergamot range 50 – 550 mls/min,
300/2	Butanol at 300 mls/min, Bergamot range 560 – 990 mls/min.

The n-Butanol has a molecular weight of 74.12, so a flow of 100mls/min is equivalent at 21° C to a concentration of approximately .05 in air. The Bergamot (Givaudan) is a complex synthetic analogue of natural bergamote oil, and has a strong citrus-*eau-de-cologne* odour at 990 mls/min; it is suprathreshold for most subjects over the lower concentrations used, and is almost always suprathreshold in the */2 ranges.

Initially a pure Bergamot was presented at a concentration of 1 l/min as being an example of an intensity of 100 arbitrary response units. Subjects were instructed on each trial to make two responses always in the same order, as soon as possible after sniffing the airstream through a flap in the side of a perspex chamber containing the odorous flow. These responses were "B" – the intensity of the odour associated with Bergamot, and "T" – the intensity of the total undifferentiated mixture, comprising all qualities of odour.

The stimulus mixtures were each presented for three seconds only at 30 second intervals. Ambient air was controlled at $21 \pm 1°C$.

Subjects

Volunteers of both sexes in the age range 17 - 27 years served. Full written consent to participate was obtained, and note was taken of any suspected abnormalities in their olfactory senses and of any extreme likes or dislikes. No screening on this basis was found necessary.

Results

Interim Checking Procedures.

Subjects were permitted to give "null-B" ($y_B = 0$) or "null-T" ($y_T = 0$) responses if they felt that they could not detect any odour, either specific in the case of "B" or generally in the case of "T". Most of these null responses were associated with weak Bergamot concentrations (< 100 ml/min) in condition 100/1. "null-B" $= B_0$ responses are about twice as likely as complete "null-T" $= T_0$ responses in 100/1, the percentage of such responses is given in Table 12.1. The minimum number of responses on which a percentage is based is $56 \times 5 = 280$.

Table 12.1

Percentages of Null Responses

Condition	$y_B = B_0$	$y_T = T_0$
100/1	22.5	11.1
100/2	1.1	0.0
300/1	4.6	2.3
300/2	2.9	0.3

In Table 12.1 it may be noted that increasing the Butanol background (from 100/1 to 300/1) decreases the probability of giving a null-B or null-T response, for weaker Bergamot concentrations, but slightly increases the effect for a stronger Bergamot range. For regression analysis the null responses were scored as 1, so that $log R = 0$. No actual detection responses of perceived intensity less than 5 were actually given by subjects. The difference of B_0; $22.5 \Longrightarrow 4.6$ is associated with $p < .05$ (ordinal runs test).

A minor check was made by regressing $(y_T - y_B)/y_T$ on U_B. The slope of this was negative in 19 out fo 22 cases and positive in 2, which is as expected if the fixed masking component u_M is associated with an implicit Y_N (where Y_N is to be read as 'response to Butanol odour'). This y_N should not increase with u_B or with y_B, unless one possible form of synergy is operating.

Static Analysis

The zero-order descriptive statistics for all the 22 *Ss* are summarized in Table 12.2a; as expected there is a considerable difference between the information transmission in the concentration-intensity relationship as the concentration range is elevated. The identifiability and discriminability of the more concentrated stimuli is considerably less than for the weaker stimuli, which is, of course, an expression of Weber's Law. The effect of the background levels, Butanol at 100 ml/min or 300 ml/min is secondary but present. Using only the statistics of the fitted linear regression of $log R/log S$ gives the false impression that very little is transmitted for the higher concentration range. We shall see from the dynamic analysis later that this is not so.

It is of heuristic value to see if the steeper slope of $log R/log S$ is associated with a "cleaner" regression trend; Figure 12.8 shows that this is weakly so in the cases of the */1 sets but not for the */2 sets. Again, one might be led to deduce that only the */1 data sets are interpretable, and that performances are degenerated into randomness as the stimulus concentration range was raised (and hence psychophysical scaling compressed). This would be a misconception which follows from only examining zero-order system parameters.

The zero-order analysis confirms that there are strong individual differences, that the task is more efficiently performed with lower Bergamot concentrations, and that increasing the background noise via the Butanol component degenerates performance slightly.

Tables 12.2a and 12.2b

(a)   Mean Zero-Order Descriptive Statistics

Group	Condition	m	%var	SDratio
A	100/1	1.39	39.79	.47
B	100/2	.96	1.60	.57
C	300/1	1.08	32.67	.45
D	300/2	.84	2.02	.56

(b)   Mean ARMA Parameter Values

Group	Condition	b_0	d_1	c_1
A	100/1	8.48 ± 4.54	4.79 ± 5.11	3.15 ± 0.15
B	100/2	1.78 ± 0.50	1.19 ± 1.27	0.57 ± 0.25
C	300/1	4.87 ± 3.41	1.54 ± 1.01	0.45 ± 0.18
D	300/2	1.59 ± 0.81	1.35 ± 1.07	0.56 ± 0.23

Notes: (1) Large b_0 differences are associated with */1 and */2 differences, (2) d_1 and c_1 are greater for group A.

Dynamic Analysis

The ARMA model to be fitted to all the data sets after filtering u_B, y_B, is

$$\hat{y}_{B_j} = (b_0 u_{B\cdot j} + b_1 u_{B\cdot j-1}$$
$$+b_2 u_{B\cdot j-2} + b_3 u_{B\cdot j-3} + b_4 u_{B\cdot j-4} + b_9 u_{\cdot j-9})/(1 - a_1)$$
$$+(d_1 e_j + d_2 e_{j-1} + d_3 e_{j-2})/(c_1 e_j + c_2 e_{j-1} + c_3 e_{j-2}) \qquad [12.7]$$

This can be fitted to all individual Ss data sets u_B, y_B and u_B, y_T with satisfactory small residuals at all lags ±12, using SCA (1985). The first-order correlations between residuals of $u_{\Phi B}$ and $\hat{y}_B$ were, in 14 cases out of 22, $\leq \pm.10$. In Table 12.3a the column headed _SDratio_ is derived after fitting the generalized $\hat{y} =\text{ARMA}(\Phi u) + \text{ARMA}(e)$ model of [1], divided by the standard deviation of the original y series. If the fitted variance is required this is $1 - (SDratio)^2$. We find that for the low concentration stimulus range 79% is taken up by the model, whereas for the high concentration range only 68% is modelled. In both cases the part modelled has an identifiable dynamic structure. The usual very wide individual differences in performance are noted, which are properly reflected in the estimates of the coefficients $\{b, c, d\}$ but generally the dominant terms as expressed in their associated t values ($t = $ estimate/s.e.) are b_0, c_1, and d_1. There is an evident tendency for b_9 to be large (checked by the auto-, partial, inverse and extended autocorrelation functions) which may be associated more readily with the 300/* conditions, as shown in Table 12.3b. The data are too sparse to explore this point further, although it makes some sense.

Table 12.3

Number of Ss with $t \geq 1.9$ at each
a, b, c, d coefficient in Eqn [12.7]

	100/* (in 10 Ss)	300/* (in 12 Ss)
b_0	8	8
b_1	2	3
b_2	3	3
b_3	2	0
b_4	2	2
b_9	1	*7
a_1	4	7
d_1	4	4
d_2	2	*7
d_3	3	6
c_1	4	5
c_2	1	*6
c_3	3	5

The transfer function form [12.1] of a time series analysis is convenient for forecasting, but here it is more useful to construct the spectrum of impulse response coefficients, given negligible residuals and no significant autoregression in the residuals. The factorization of this spectrum, treated as a polynomial in z^{-1} (Gregson, 1983, 1984) leads to a representation which is closer to dynamic systems theory and facilitates comparison with our previous related findings on 2-input psychophysical olfactory tasks. Given the condition that all roots lie outside the unit circle, there exists an infinite series representation of the system [12.1] in the form:

$$y_j = v_0 u_j + v_1 u_{j-1} + v_2 u_{j-2} + \ldots \ldots \qquad [12.8]$$

If $A(z^{-1}) = 1$ then this series in $\{v, u\}$ is finite, but in our data this condition is clearly violated in at least half the data sets. Using a truncation at v_9 (10 terms) and factoring in the $\sigma + i\omega$ plane (so that the lefthand halfplane represents stable convergent dynamic components), we can readily identify the dominant conjugate or real roots outside the unit circle. It is already known that psychophysical data of the sort met here commonly include one or two conjugate pairs (oscillating components) and possibly a real negative root (meaning stable and convergent transfer of input to output). Instability, (and limit cycles) shows when some roots cross over into the righthand halfplane. In the terminology of Robinson (1983) psychophysical systems are mixed and not always minimum delay, but secondary roots with modulus less than unity have been ignored in the analyses comparing the four groups used here.

The fitted models were used in all cases to compute the first 10 v_k, the Impulse Response Coefficients (i.r.c.s). Both in terms of the psychophysics and statistically this is sufficient; long tail terms are both increasingly miniscule and uninterpretable. Visual examination of the obtained results suggested that for the purpose of checking differences between the four stimulus conditions only the first five i.r.c.s were sensible input for a discriminant analysis. Such an analysis is in this context an iterative strategy, as can be shown by using it where necessary recursively instead of as a single-pass computation.

Table 12.4

First Five I.R.C.s for Each Group

	v_0	v_1	v_2	v_3	v_4
100/1	.131±.049	.016±.006	-.003±.045	-.021±.025	-.006±.029
100/2	.045±.022	-.002±.025	-.010±.025	-.038±.023	-.049±.019
300/1	.100±.062	-.003±.070	-.013±.038	-.024±.018	-.001±.032
300/2	.022±.049	.012±.040	.011±.021	-.008±.047	.009±.059

Notes: (1) Large v_1 differences associated with */1, */2 differences. (2) v_3, v_4 differences associated with 100/2, 300/2 differences.

Table 12.4 gives mean values of v_0 through v_4 (the i.r.c.s) for each group. This curtailed representation is sufficient to illustrate the major differences associated with the four conditions, and as input to a discriminant analysis yields a separation on the first two canonical variables shown in Figure 12.9. This configuration is primarily *dependent upon* v_0 and v_4, it is the result of assuming that each response pattern is the consequence of a specific stimulus condition; that is, the four groups generate specific dynamic response signatures. This assumption is neither necessary nor strictly plausible, because the Ss have each a diversity of possible response patterns which they might employ, and the relative probabilities of being induced to function within one such pattern are linked to the stimulus condition but only stochastically. Then we should expect that some response patterns appear more readily within one group than within another, but that the most effective discriminant representation is in terms of response patterns and not in terms of stimulus conditions *per se*. Reallocation of 9 Ss to the groups in which the discriminant analysis underlying Figure 12.9 placed them immediately yields a new clustering with 100% effective allocation.

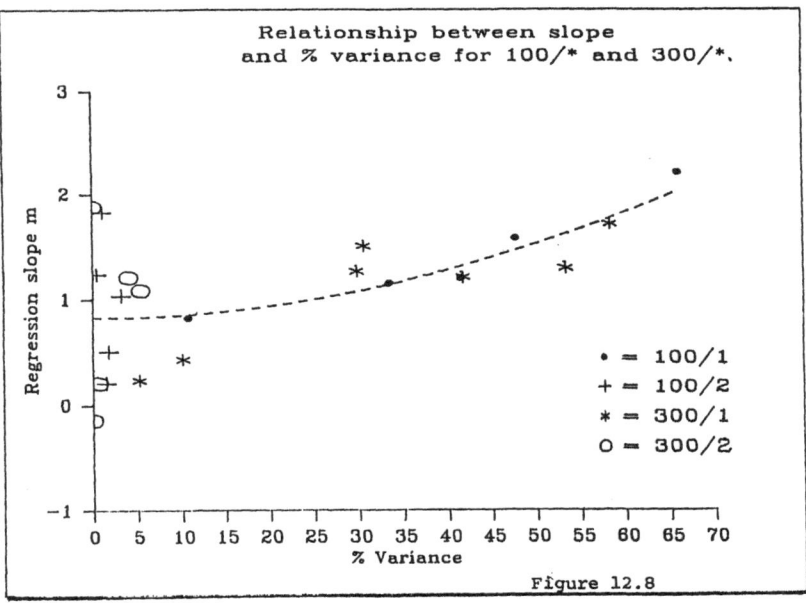

Figure 12.8

The groups are now:

$$A' = \text{mostly } 100/1$$
$$B' = \text{mostly } 100/2$$

and two remaining clusters which can be treated, from inspection of Figure 12.10, as one intermediate case. The separation is very strong and the canonical variable #1 (X) separates A' from B' and may be interpreted as the shift in i.r.c.s as a function of the stimulus range; changing from low to high induces a shift from A/ to B/ in some subjects.

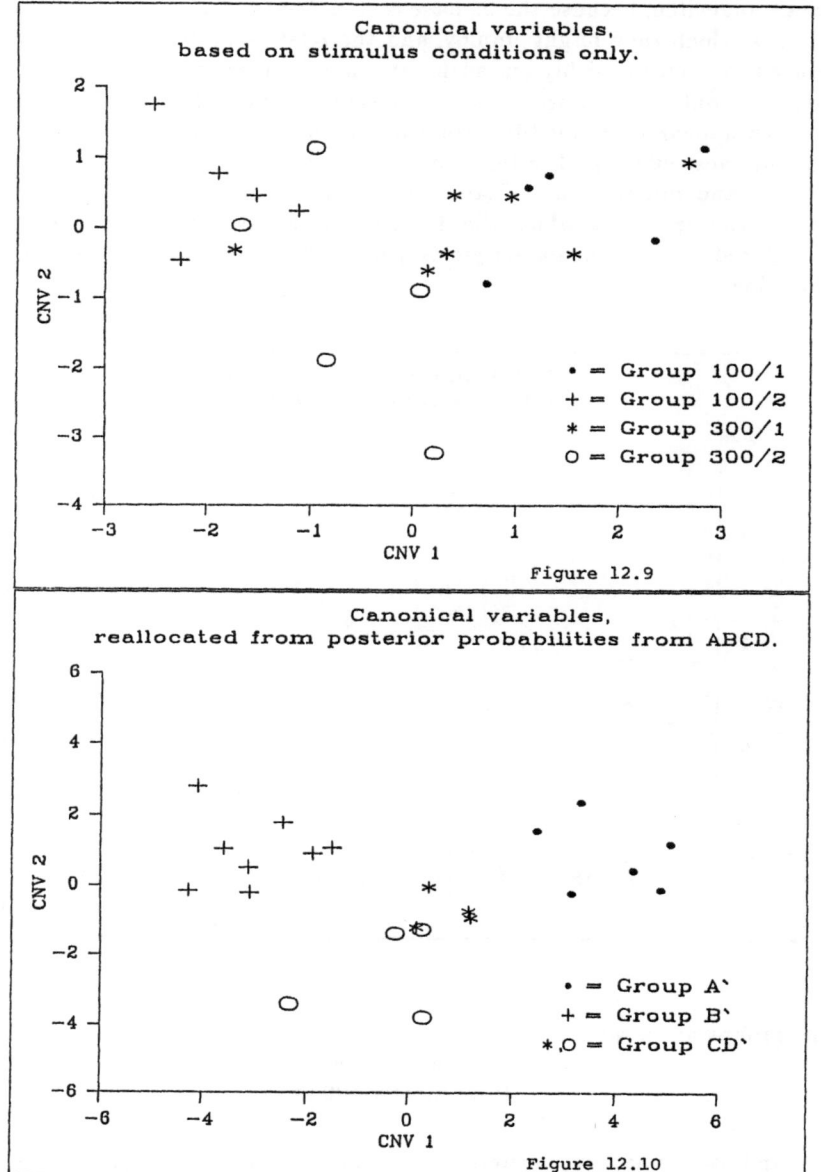

Figure 12.9

Figure 12.10

To make better sense of the system dynamics which are partly revealed by the clustering of Figure 12.10, we return to the i.r.c.s in the full 10 lags form as computed, and factorize these as polynomials in the z shift operator. This yields distinct topographies in the $\sigma + i\omega$ plane for each cluster. Table 12.5 lists those roots which are the largest outside the unit circle (moduli 1), some slight rearrangement to put C1 = largest complex root, C2 = complex root nearest the unit circle, R1 = largest real root, and R2 = a second real root of importance, facilitates comparison.

It is seen that A' includes two complex roots in the lefthand (stable) halfplane, and no other regular features. The loci of the C1 roots are shown in Figure 12.11 and indicate that the rapidity of convergence increases with the real part of the complex root. The trajectory does not cross over into the unstable righthand halfplane. This pattern of responding, which implicates two feedback processes coupled (but not necessarily linked to the two sensory components of the inputs, as the analysis here is only in terms of Bergamot odour) may be a consequence of both the signal and the response noise being separately autoregressive.

Subgroup B', as shown in Figure 12.12, mainly arising under the conditions of group 100/2, shows one or two conjugate roots and always a positive real root which is divergent but dominated by the complex conjugate oscillating roots in the lefthand halfplane. These real roots are unstable and if the corresponding T series of responses are likewise analysed may then show a flip over into the opposite sign. In one case (# 22) we have for consistency used the real roots in the T analysis; in all cases the complex roots are neighbours in both B and T series. Hence, the degradation of performance contingent upon either the high concentration series in 100/2 or perhaps due to poor discrimination in the observers in the */1 cases results in a divergent nonoscillating component which carries little information and is effectively noise. The complex root trajectory runs in the right hand halfplane, and thus has quite different dynamics from the A' group. Crossing the $i\omega$ axis suggests that the process is going into limit cycles intermittently.

The third cluster CD' we have identifed from Figure 12.10 is weak perception with a negative real root. It may be that this case is one in which the information that the system transmits is carried by a real root dominating the complex ones; the real negative root transmits a strong signal due to intense stimulation but it is undifferentiated with respect to intensity, the capacity for signal differentiation rests only with the largest complex roots. T roots have been used in two cases.

To explore this problem of system dynamics further we need the results summarised in Tables 12.6a and 12.60b, giving the coefficients of a canonical correlation analysis linking the coordinates of Figure 12.10 to the zero-

Table 12.5
Summary of major roots outside the unit circle
of i.r.c. factorization of B or near T responses
for subjects reallocated to maximally discriminable
subgroups, based on a discriminant function of the
first five i.r.c.s as in Table 12.4.

Original S #	C1	C2	R1	R2
Subgroup A'				
(resembles 100/1)				
1	-3.41±4.51	-.54±1.60	-	-
3	-.65±2.48	-.07±1.11	-34.01	
6	-2.35±3.90	-.32±1.48	-	-
8	-3.63±1.91	-.71±1.18	-	-
18	-3.50±5.16	-.57±1.67	-	-
19	-.66±2.37	-.14±1.07	-9.78	-
Subgroup B'				
(resembles 100/2)				
2	-	-.38±1.61	+7.28	-10.42
4	.37±3.16	.21±1.38	+5.35	-
7	-1.73±1.08	-	+5.74	-1.00
9	-	-.28±1.52	+9.49	-3.03
11	-1.06±2.76	-	+11.39	-
13	-.25±2.39	-.02±1.05	+7.88	-
17	.25±2.70	.24±1.07	+3.12	-
22	-	-.74±1.17	T+4.12	T-3.09
Subgroup CD'				
(300/*)				
5	-5.06±7.24	-.51±1.61	-5.98	-
10	-.57±2.07	-.22±.99	-7.96	+1.48
12	1.20±12.82	-.15±1.26	T-5.83	-
14	-1.05±2.42	-.36±1.03	T-36.05	-
15	-.92±2.15	-.68±1.01	-3.08	-
16	-.12±2.58	.02±1.08	-21.54	-
20	-.84±2.89	-.15±1.32	-11.59	-
21	-	-.09±1.54	-4.58	+3.66
Γ theoretical simulations				
low2	-2.73±2.03	-.80±1.44	-	-
low6	-2.77±1.88	-.81±1.40	-	-
high2	-1.39±1.89	-.52±1.07	-4.41	-
high6	-1.25±1.89	-.49±1.07	-5.69	-

order statistics of Table 12.1 and some measures of the root moduli (but signed for their respective halfplanes) from Table 12.5. It is informative to perform the canonical analysis twice, once for all the data and once excluding the relatively uninterpretable CD' group. From Table 12.6a we can see that the strongest first-order linkage is between X and %variance; that is the separation of clusters relates in the major canonical axis to the efficiency of transmission as a linear S$\Longrightarrow$R mapping. Only CNV1 is strictly interpretable.

In Table 12.6b again only one canonical variable deserves attention, CNV1, and again dimension X is linked to both the %variance and to the SM variable, which is the sum of the moduli of the complex roots, of which C1 clearly dominates. This finding, linking information transmission with the modulus of the largest complex root, is compatible with a related finding by Gregson and Gates (1985). The second root CNV2, though not associated with a sufficiently low probability, deserves notice because it suggests, taken with CNV1, that differential sensitivity (m) in the intensity judgment task maps into different dynamic roots from those mediating the response trend line. In an earlier study on odour mixtures (Gregson, 1982, 1983a) it was noted that transmission of odour quality (necessary to separate out B and T judgments here) and transmission of odour intensity undifferentiated with respect to type appeared to implicate quite separate dynamic roots. The present results support that conjecture. It follows that the i.r.c. spectrum for T judgments need not have exactly the same factorization as that for B judgments, as used so far, though because the T stimuli are all of the form *B plus a constant* unless the system has some strong nonlinearities the dynamic model will be almost the same for B and for T. It is the nonlinearities which have been traditionally treated as intercomponent masking or more rarely intercomponent facilitation, which might show in comparisons of the root loci of the two response sequences. These however on examination appear to be almost random and in the case of the conjugate roots very small. It does not follow that this result would hold if the second component were of mean perceived intensity comparable to that of the Bergamot. The earlier results of Gregson (1984a) suggest that this would increase the dynamic complexity of the situation.

The root locus configurations for A'and B' are also graphed in Figs 12.11 and 12.12 respectively. It is seen that the larger conjugate roots can be treated as lying on trajectories which are very different in the two cases A' and B'; in Figure 12.11 the trend is for more rapid convergence (rate of convergence is proportional to ω^{-1}) to be associated with larger, dominant $| \sigma |$, but the reverse trend occurs in Figure 12.12 and the trajectories cross over into the unstable righthand halfplane, suggesting that the pro-

Tables 12.6a and 12.6b
Canonical Correlation Analysis
Relating X, Y Coordinates in the Discriminant Analysis
yielding A', B', CD' to
the signed moduli of C1, C2, and R1 in Table 12.8
SM $= |C1| + |C2|$, SqCC $=$ Squared Canonical Correlation

(a)   Using the full data set of 22 Ss

	Standardized Coefficients	
	CNV1	CNV2
X	.980	-.201
Y	.201	.980
m	.101	-.853
%var	.737	1.114
sgn$|R1|$	-.165	-.780
sgn$|C1|$	-.085	.044
sgn$|C2|$	-.255	.147
SM	.198	.089
SqCC	*.736	.089

(b)   Using only the data sets of 14 Ss, from A' and B/.

	Standardized Coefficients	
	CNV1	CNV2
X	1.003	-.010
Y	.089	.999
m	-.005	-1.389
%var	.482	1.925
sgn$|R1|$	-.165	-.896
sgn$|C1|$	-.497	.103
sgn$|C2|$	-.078	.462
SM	.548	-.943
SqCC	*.818	.619

$^*p < .05$

cess is now forced into limit cycling by the higher concentration inputs. This would be a dynamic consequence of fluctuations induced by quicker adaptation and recovery with a quasiperiodicity induced by the temporal forcing function of the trial spacings at half-minute intervals.

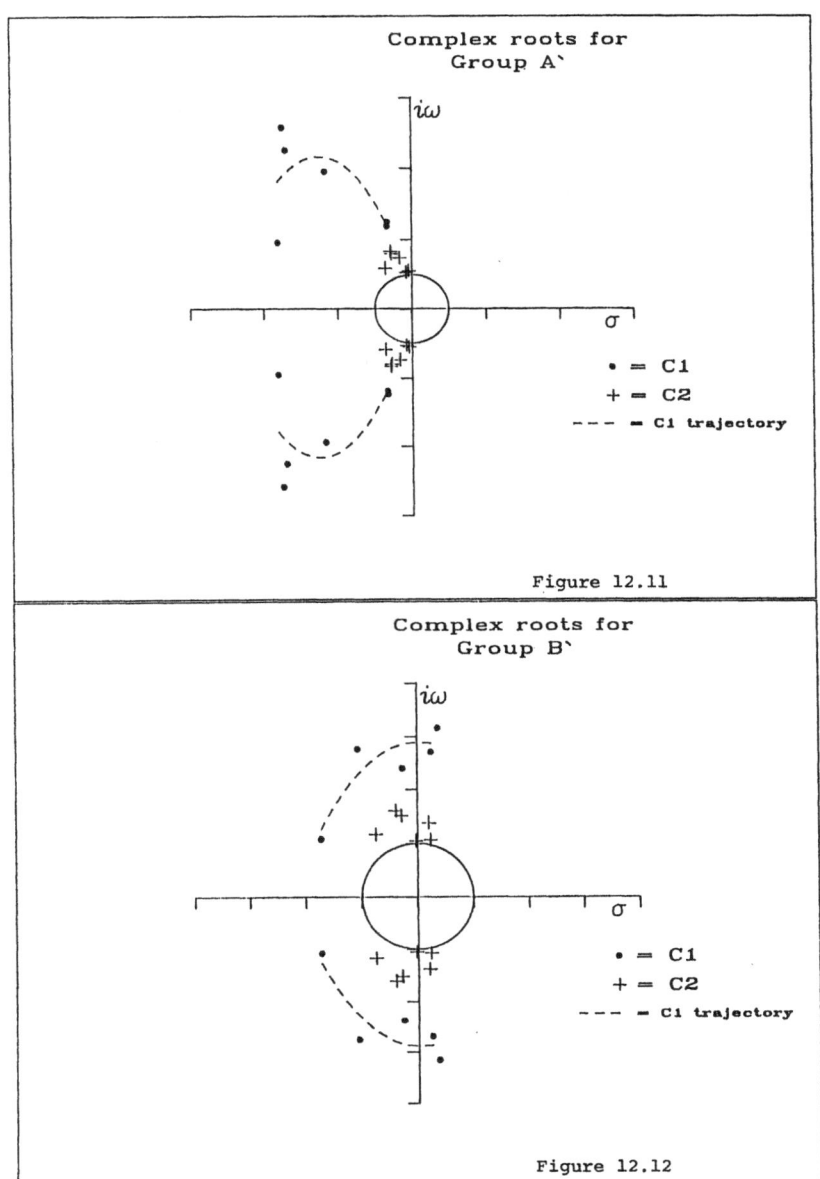

Figure 12.11

Figure 12.12

It may be concluded that the dynamics of sequential intensity (or magnitude) estimation vary markedly with odour stimulus levels, and that individuals may function in one of a number of modes, depending both on the sensory input and on the individual's capacity to process information that is both noisy and protracted in time. The dynamics of sequential effects may perhaps be modelled by one generic equation of sufficiently high order, but the variation of the parameters within that equation reflect a wide range of psychophysical behaviour processes which to the external observer would be disparate. There is some neurophysiological basis for a close link between temporal sequential processing and sensory input, via the prefrontal cortex. Fuster (1981, pp. 134 - 136) advances the view that sensory information processing involves *'the temporal integration function of the prefrontal cortex.....a form of memory that permits referring any event in a behavioural sequence to preceding events.'* It would follow that in the sort of task used here both the peripheral (receptor) sensory adaptation-recovery cycle and the cortical (dorsolateral convexity) integration are implicated and to some extent confounded in their effects on observable performance.

The comparison to be drawn here is between a low $2.5 \leq a \leq 3.2$ (or high $3.2 \leq a \leq 3.9$) range of a, with $e = .15 \pm$ noise, $\eta = 8 \pm$ noise, and the lower (or higher) Bergamot concentration ranges respectively. The output pair of series $\{U\}, \{Y_{ob_s} \mid a, e, \eta\}, J = 1, 2, ..., N$ where U has some rectangular i.i.d. form (as has u_B in the odour experiment here) can be modelled similarly to $\{u_B\}, \{y_B\}$, via [12.1], and consequently by [12.7], [12.8], giving the root loci listed under the cases Γ low2, low6, high2, high6 in Table 12.5. The variants 2, 6 after low, high denote measures of constraint of $\Delta^1 Y_j$ imposed as a boundary condition within the recursive loop of Γ. They are not of importance here, but less constraint induces as expected a change in the autoregressive structure of Y_{ob_s}. It is of interest here that the negative real root found in CD' is parallelled in the high a range data from Γ V 7; the dynamics induced by system nonlinearity in Γ resemble the effects of high background noise in the real data.

References

Ahlström, R. and Baird, J. C. (1987) Shift in stimulus range and the exponent of the power function for loudness. *Reports of the Department of Psychology, University of Stockholm*, No. 653.

Bennett, R. J. (1979) *Spatial Time Series* , London: Pion.

Box, G. E. P. and Jenkins, G. M. (1970) *Time-Series Analysis; Forecasting and Control.* San Francisco: Holden Day.

Cain, W. S. (1975) Odor intensity: Mixtures and masking. *Chemical Senses and Flavor, 1*, 339 - 352.

Cain, W. S. and Drexler, M. (1974) Scope and evaluation of odor counteraction and masking. *Annals of the New York Academy of Sciences* , *237* , 427 - 439.

Crozier, W. J., and Holway, A. H. (1937) On the law for minimum discrimination of intensities. *Proceedings of the National Academy of Sciences, 23,* 23 - 28.

Freeman, W. J. and Viana di Prisco, G. (1984) EEG Spatial Pattern Differences with Discriminated Odors: Manifest Chaotic and Limit Cycle Attractors in Olfactory Bulb of Rabbits. *In* Palm, G. and Aertsen, A. (Eds.) *Brain Theory.* Berlin: Springer-Verlag.

Fuster, J. M. (1981) *The Prefrontal Cortex.* New York: Raven Press.

Gregson, R. A. M. (1982) Representation of a 2-input 2-output odour mixture identification task as a multivariate time series. *Proceedings of the Second Australian Mathematical Psychology Conference,* Newcastle, N.S.W., Newcastle University Press.

Gregson, R. A. M. (1983a) The sequential structure of odour mixture component intensity judgments. *British Journal of Mathematical and Statistical Psychology, 36* , 132 - 144.

Gregson, R. A. M. (1983b) *Time Series in Psychology.* New Jersey: L. Erlbaum Associates.

Gregson, R. A. M. (1984a) Invariance in Time Series Representations of 2-input 2-output Psychophysical Experiments. *British Journal of Mathematical and Statistical Psychology, 37,* 100 - 121.

Gregson, R. A. M. (1984b) Behaviour of a system with gain and pure delay filters incorporating a nonlinear difference feedback loop as a generalized psychophysical model. Seminar, Humboldt University, Berlin, DDR.

Gregson, R. A. M. (1984c) Similarities between odor mixtures with known components. *Perception and Psychophysics, 35,* 33 - 40.

Gregson, R. A. M. (1985) The Subjective Weber Function and Output Uncertainty in Nonlinear Psychophysics. Paper presented at *Osnabrück Mathematical Psychology Meeting 1985* , Universität Osnabrück, FRG.

Gregson, R. A. M. (1986) Qualitative and Aqualitative Components of Odour Mixtures, *Chemical Senses, 11,* (in press).

Gregson, R. A. M. and Gates, A. (1985) Cross-modal Identification: Effects of Contingent Changes in the Stimulus Series. *Biological Cybernetics, 52,* 247 - 258.

Helson, H. (1947) Adaptation level as a frame of reference for prediction of psychophysical data. *American Journal of Psychology, 60,* 1 - 29.

Lehky, S.R. (1983) A Model of Binocular Brightness and Binaural Loudness Perception in Humans with General Applications to Nonlinear Summation of Sensory Inputs. *Biological Cybernetics, 49,* 89 - 97.

Nicolis, J. S., Mayer-Kress, G., and Haubs, G. (1983) Non-uniform Chaotic

Dynamics with Implications to Information Processing. *Zeitschrift für Naturforschung, 38a,* 1157 - 1169.

Parducci, A. (1984) Perceptual and judgmental relativity, *In* Sarris, V., and Parducci, A. (Eds.) *Perspectives in Psychological Experimentation : Toward the Year 2000.* Hillsdale, New Jersey: L. Erlbaum Associates.

Prigogine, I., and Stengers, J. (1985) *Order out of Chaos.* London: Flamingo.

Robinson, E. A. (1983) *Multichannel Time Series Analysis with Digital Computer Programs. (2nd Edn.)* Houston, Texas: Goose Pond Press.

Sarris, V., and Zoeke, B. (1985) Tests of a quantitative frame-of-reference model: Practice effects in psychophysical judgments with different age groups. *In* D'Ydewalle, G. (Ed.) *Cognition, Information Processing, and Motivation.* Amsterdam: Elsevier Science Publishers.

SCA System. (1985) *The SCA System for Univariate-Multivariate Time Series and General Statistical Analysis (1985 ed.)* De Kalb, Illinois: Scientific Computing Associates.

Triesman, M. (1985) The Magical Number Seven and Some Other Findings of Category Scaling: Properties of a Model for Absolute Judgment. *Journal of Mathematical Psychology, 29,* 175 - 230.

Wilson, B., Mackintosh, N. J., and Boakes, R. A. (1985) Transfer of Relational Rules in Matching and Oddity Learning by Pigeons and Corvids. *Quarterly Journal of Experimental Psychology, 37B,* 313 - 332.

Young, P. (1984) *Recursive Estimation and Time Series Analysis.* Berlin: Springer Verlag.

Zeeman, E. C. (1976) Duffing's Equation in Brain Modelling.*Institute of Mathematics and its Applications Bulletin,* 207 - 214.

Zoeke, B., and Sarris, V. (1983) A comparison of "frame-of-reference" paradigms in human and animal psychophysics. *In* Geissler, H. - G., Buffart, H. F. J. M., Leeuwenberg, E. L. j., and Sarris, V. (Eds.) *Modern Issues in Perception.* Berlin: VEB Deutscher Verlag der Wissenschaften. Amsterdam: North Holland.

13 Mixing and Attenuation of Sensory Dimensions

The motivation for considering what seems to be a general problem in a very circumscribed way, as is to be done here, arises from results reported by Gregson (1986) on the problem of hypoadditivity in olfactory psychophysics. *Hypoadditivity* is a phenomenon in which the intensity of a total mixture is in some sense less than the sum of the component intensities experienced separately. Some background is helpful, as is the use of a notation appropriate to draw distinctions between responses to a single stimulus in isolation, and to the same stimulus when it is a component of a mixture, or is partly masked by system noise which is permanently present.

The latter idea is, of course, central in SDT (see Chapter 11) but there is in olfaction some neurophysiological evidence involved (Freeman and Viana di Prisco, 1984). The representation of sensory intensity in the olfactory bulb appears to implicate two oscillating activity patterns, one associated with the resting state, and the other with stimulus input. When two such oscillators couple, limit cycling can be produced, and the phase of the cycle can be the basis for encoding sensory intensity. The cycling in not linked to a fixed tissue pathway, but meanders through continually changing loci. At the same time, different stimulus qualities generate different excitation patterns, and these patterns can be formed *de novo* (Baird, 1986). There is thus a potential extension of the type of theorizing advanced here, into considerations of pattern, or rather quality, learning and recognition. To make such an extension requires additional assumptions outside the scope of these chapters, but neurophysiologically it is not unrealistic.

This neurophysiological evidence is pertinent to the central processing of a single input; a pure substance, generating one qualitative sensation, or at low intensity input a sensation without clearly describable quality, that is, an aqualitative sensation. Heuristically one might link this minimal resting-state activity with a_{min} in Γ, and then a range of qualitative intensities with a values above some lower bound.

Let us define a mixture set of two substances at chemical concentrations u_1 and u_2 as the three possible presentations in a carrier medium, usually air and sometimes nitrogen, as

$$| \, u_1 \quad u_2 \, | \qquad | \, u_1 \quad 0 \, | \qquad | \, 0 \quad u_2 \, |$$

and further require that psychologically the u are suprathreshold and qualitatively discriminable, and chemically they are not near saturation level. The u are in physical units, and the corresponding responses will be estimates of intensity on some rating scale Y, y. We consider only perceived intensity ratings, which we refer to as judgments and not identifications of odour quality changes. Three different responses must be distinguished:

Y, the overall intensity of $u_1 \odot u_2$, where $\odot$ means co-presentation simultaneously of the two odourants as in $| \, u_1 \quad u_2 \, |$,

$_m y_1$, $_m y_2$, the two odour component judgments of u_1 in $u_1 \odot u_2$ and u_2 in $u_1 \odot u_2$ respectively,

y_1, y_2 the judgments of the (pure) component odours u_1 and u_2 at concentrations corresponding to those in the mixture $| \, u_1 \quad u_2 \, |$.

We could write

$$Y = \xi_1(u_1 \odot u_2), \qquad _m y_1 = \xi_2(u_1; u_1 \odot u_2), \qquad y_1 = \xi_3(u_1),$$

and so on, noting then that the mathematical problem is that we do not know if the functions ξ_1, ξ_2, ξ_3 are the same for given u_1, u_2. In other words, the odour mixture problem arises because there are difficulties, empirically, in predicting one of the Y, $_m y_i, y_i, i = 1, 2$ measures if we only know one or two of the others.

The psychophysical evidence is not unequivocal, though in general it supports findings that

$$Y < \Sigma y_i$$

$$_m y_i < y_i \qquad\qquad\qquad [13.1]$$

$$\text{and hence } \Sigma_m y_i < \Sigma y_i,$$

with an indeterminate relationship $Y \gtrless \Sigma_m y_i$.

Let us consider only the case where there are two dimensions, but endeavour to formulate assumptions so that the extension to k dimensions should be obvious, by adding some more terms linearly and taking maxima over sets of k alternatives when appropriate.

If there are two stimulus inputs, i, i' for example, of Γ V 7 type, each with parameters $\{a_{min}, e, (\Delta^1 a)_{max}, \eta\}$, denoted by $\Gamma(U_i)$ and $\Gamma(U'_i)$, and the associated noise levels are characterised sufficiently by

$$a^* \sim a_{min} + \delta a, \qquad \delta a \sim \mathbf{N}(0, \sigma_a),$$

$$e^* \sim e + \delta e, \qquad \delta e \sim \mathbf{N}(0, \sigma_e),$$

and

$$\eta^* \sim \eta + \delta\eta, \qquad \delta\eta \sim \mathbf{N}(0, \sigma_\eta),$$

where all of σ_a, σ_e, and σ_η are second-order with regard to their respective associated parameters; this gives

$$Yb_i = \Gamma(Ub_i) = \Gamma(a^*, e^*, \eta^*)_i \qquad [13.2]$$

Suppose that we adopt the following:

def 13.1
$$Yb_{ii'} = \max(Yb_i, Yb_{i'})$$

def 13.2
$$y_i = \Gamma(U_i) - Yb_i$$

def 13.3
$$_m y_i = \Gamma(U_i) - Yb_{ii'}$$

def 13.4
$$Y = \Gamma(U_i) + \Gamma(U_{i'}) - Yb_{ii'}.$$

The definition **13.4** leads to predictions that are false. It follows from **13.1**, **13.2**, and **13.4** that

$$\Sigma_m y_i - Y = -Yb_i - Yb_{i'} + Yb_{ii'} \qquad [13.3]$$

so

$$Y - \Sigma_m y_i = \min(Yb_i, Yb_{i'}) \qquad [13.4]$$

and

$$Y - \Sigma_m y_i = +2.Yb_{ii'} - Yb_{ii'} = Yb_{ii'} \qquad [13.5]$$

or

$$Y > \Sigma y_i > \Sigma_m y_i \qquad [13.6]$$

This does not match [13.1]; the expression which gives trouble is **def 13.4**. As has been commented earlier, to assume that the system can combine two input pathways is to add some other combinatorial operation downstream from our simple recursive one-dimensional loops. The separate outputs of two loops have to be summed unless we postulate additive operations even *before* the entry into the unidimensional loops.

Suppose that cascading, already described in Chapter 7, is the basis for addition. Then it follows that

def 13.5
$$Y_c = \Gamma(Y_i(\text{Re}) + Y_{i'}(\text{Re}) - Y_{bii'}(\text{Re}))$$

How this Y_c will behave relative to Σy_i depends on the parameter sets at both stages in the cascade. The case where **def 13.4** is used to generate $Y_{,m}$ y_i and y_i has been called Γ V 10 and the process of reducing $\Gamma(U)$ by $\Gamma(U_b)$ called *attenuation*.

Some simulation results are shown in Figures 13.1 to 13.4. The relative variance in the noise process in Figures 13.2 and 13.3 is arbitrary. The interesting consequence is the breaks in Figures 13.4. This partly depends on the diminution of the information content in the cascading operation. This is not implausible as a model of real behaviour, because it is often observed, for example, that intensity judgments for odour mixtures are associated with low information transmission, only three response levels being effectively used.

Table 13.1

Output Information values associated with
a Γ V 10 and Cascading simulation
(values correspond to some of the graphs 13.1 to 13.4)

$$U_i = \Gamma_{V7}(3.2, .12, 10) \qquad U_{i'} = \Gamma_{V7}(2.9, .14, 10)$$

(max information based on a rectangular input U distribution)

Case	Info (**H**)	Propn. Max**H**
$Y_i(\mathrm{Re})$	2.164	.6332
$Y_i(\mathrm{Im})$	1.588	.4779
$Y_{i'}(\mathrm{Re})$	2.606	.7843
$Y_{i'}(\mathrm{Im})$	1.727	.5198
$Yb_i(\mathrm{Re})$	.758	.228
$Yb_i(\mathrm{Im})$	2.765	.8324
$Yb_{i'}(\mathrm{Re})$	.663	.1995
$Yb_{i'}(\mathrm{Im})$	1.955	.5886
$Y(\mathrm{Re})[\mathbf{13.4}]$	2.625	.7903
$Yc(\mathrm{Re})[\mathbf{13.5}]$	1.633	.4915
$Yc(\mathrm{Im})[\mathbf{13.5}]$	.905	.2726

Negative Masking

The phenomenon called *Negative Masking* (the name appears to be due to Laming (1986)) arises at very low stimulus intensities. It is a consistent violation of Weber's Law, in that no longer is it the case that

$$\frac{\Delta S}{S} = k_w \qquad [13.7]$$

because the "constant" k_w shows first a monotonic fall over the stimulus range, with decreasing input, and then near to zero it rises again. Laming catalogues with illustrations a number of well-substantiated cases in a diversity of sensory modalities, so that typically the graph of stimulus amplitude (U on the x-axis) against threshold amplitude (ΔU on the y-axis) is in the shape of a square root sign, thus:

The relative proportions and shape of such graphs are of course variable with the stimulus ranges and scaling procedures used. This situation suggests, if we are committed to the methodology of (traditional) stochastic

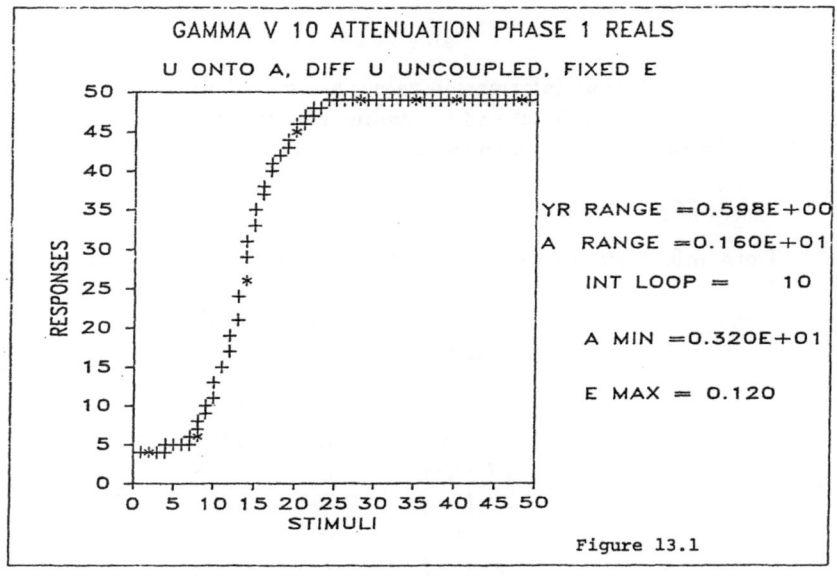

Figure 13.1

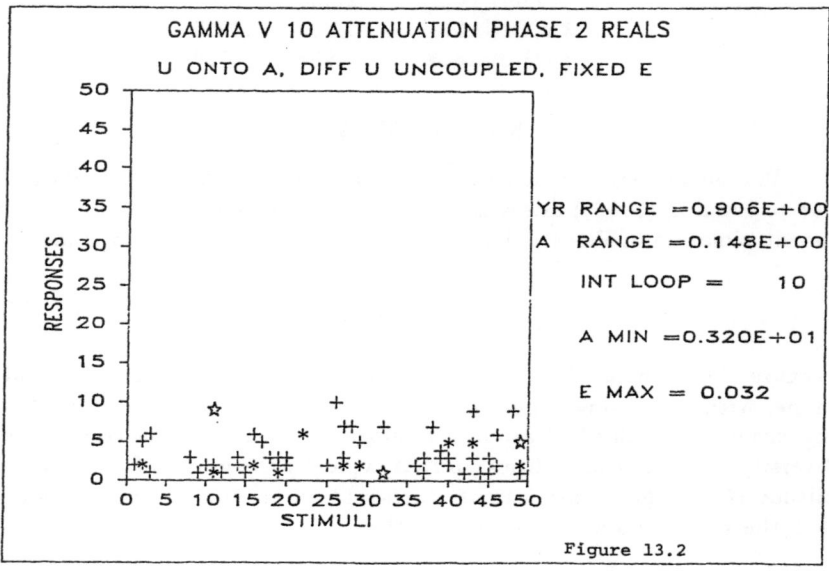

Figure 13.2

modelling in psychophysics, that the negative masking arises because the background noise becomes dominant in the near-threshold region for detection, and thus noise swamps effects due to signal properties [1]. The algebraic derivation of model structure to fit negative masking becomes somewhat laborious, with ancillary assumptions.

An analogous sort of masking follows for Γ if local variation of output is assumed to be generated by a weighted mixture of $Y(\text{Im})$ and $varY(\text{Re})$ components, under some reasonable constraints on $\{a, e, \eta\}$. That is, the phenomenon is reproducible but only for very low a values. Proceeding as in Chapter 4 the variance on $Y(\text{Re})$ due to local instability (in time) of $\{a, e, \eta\}$ for $\bar{\eta} = 10$, $2.05 < a < 3.05$, $\bar{e} = .15$ is generated, and at the same time $Y(\text{Im})$ over the same a range is stored. The absolute values of $Y(\text{Re})$ and $Y(\text{Im})$ are different by a factor of about 10^6, so $Y_R = Y(\text{Im})/Y(\text{Re})$, a noise/signal ratio of a sort, has to be multiplied by a scale factor c. Y_R represents a source of imprecision in the discriminability of U levels.

Putting

$$\delta Y_a = \left(\frac{a}{a_{max}}\right) \cdot w \cdot varY_a(\text{Re}) + \left(1 - \left(\frac{a}{a_{max}}\right) \cdot w\right) \cdot Y_{a.R} \qquad [13.8]$$

and plotting $\delta Y/U$ gives somewhat noisy graphs with the negative masking form, over a diversity of parameter values, as illustrated in Figures 13.5 to 13.10.

The results depend critically on the choice of c and w, with $0 < w < 1$, c has to be adjusted to make the numerical ranges of $Y(\text{Re})$ and Y_R comparable. In [13.8] a has been used as a subscript to the corresponding Y values in Γ V 7.

The assumption in [13.8] is that the relative contribution, towards externally-observable output variance, of the internal noise in $\Gamma(\text{Im})$ and the instability in $\Gamma(\text{Re})$ shifts, linearly, with the value of the U input, and as U increases the internal noise plays proportionately less and less part. For values of $w \geq .4$, $10 < \eta < 20$, the phenomenon appears to be reproducible; it diminishes and is lost in an irregular pattern if η is less. The contribution of the two components can be illustrated by setting w at .05 (Figure 13.5) and .95 (Figure 13.10) for comparison. The plots in Figures 13.5 to 13.10 are rescaled to display the topologies of the two

---

[1] An early study by Stücker (1908) compared the form of the deviations from Weber's "Law" in two groups of subjects; nonmusical people, and members of the Imperial Viennese Court Opera Orchestra under the direction of Gustav Mahler. The musicians did not show deviations at the extremities of the sensory range to anything like the same extent as did the nonmusical people.

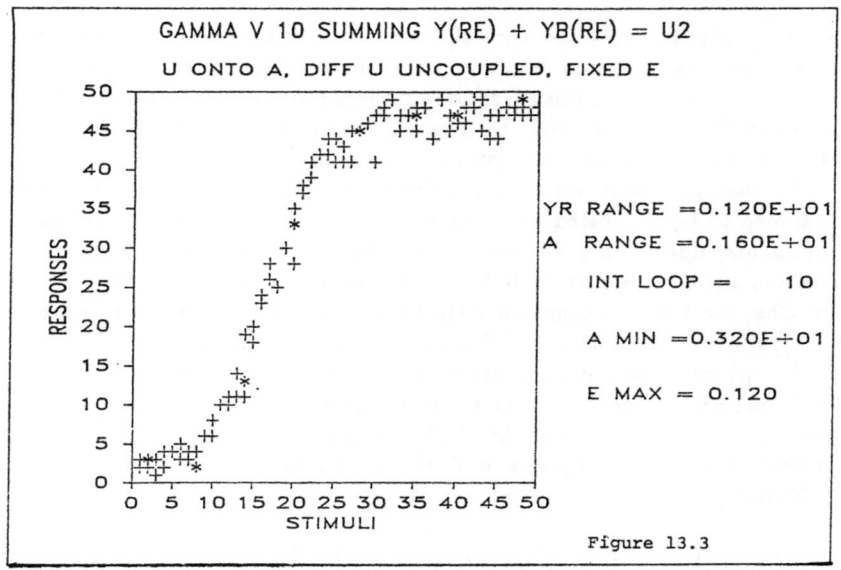

Figure 13.3

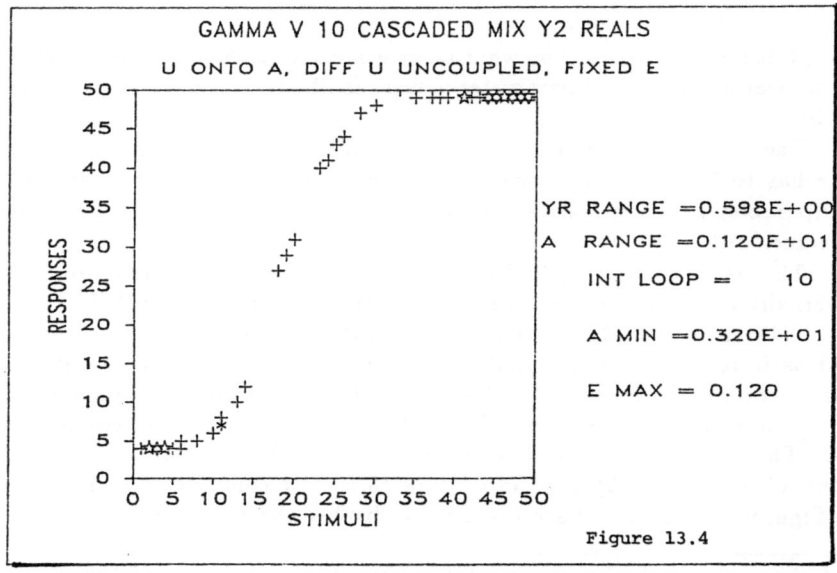

Figure 13.4

components. The interest in this phenomenon, which can only be found in real data with careful experimentation in the near-threshold region, lies in the fact that the analogy of negative masking can be produced with only one additional assumption to the Γ model, namely [13.8], and that it only appears for very low a values, below the range for which most of the simulations in this monograph are run.

References

Baird, B. (1986) Nonlinear dynamics of pattern formation and pattern recognition in the rabbit olfactory bulb. *Physica, 22D,* 150 - 175.

Cain, W. S. (1975) Odor intensity: Mixtures and masking. *Chemical Senses and Flavor, 1,* 339 - 352.

Freeman, W. J. and Viana di Prisco, G. (1984) EEG Spatial Pattern Differences with Discriminated Odors: Manifest Chaotic and Limit Cycle Attractors in Olfactory Bulb of Rabbits. *In* Palm, G. and Aertsen, A. (Eds.) *Brain Theory.* Berlin: Springer-Verlag.

Gregson, R. A. M. (1986) Qualitative and Aqualitative Components of Odour Mixtures, *Chemical Senses, 11,* 455 - 470.

Laming, D. (1986) *Sensory Analysis.* London: Academic Press.

Stücker, N. (1908) Über die Unterschiedsempfindlichkeit für Tonhöhen in verschiedenen Tonregionen. *Zeitschrift für Sinnesphysiologie,* 392 - 408.

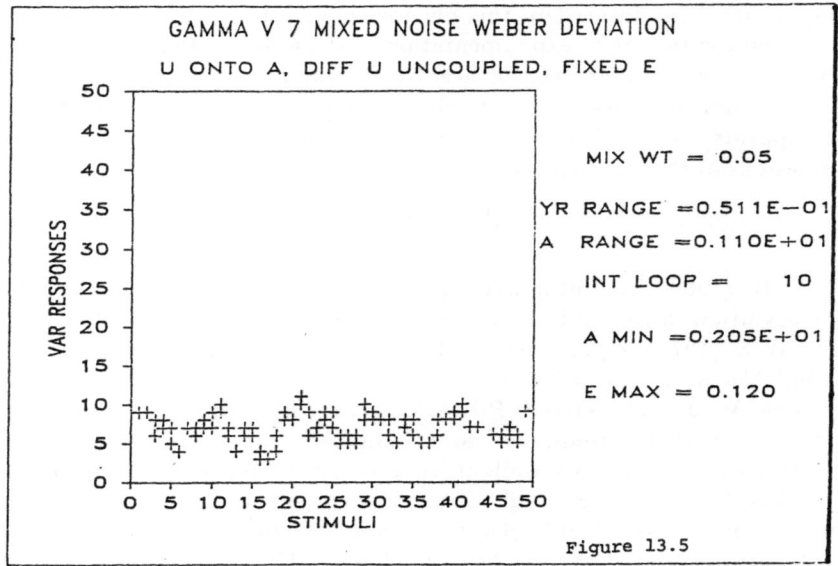

GAMMA V 7 MIXED NOISE WEBER DEVIATION

U ONTO A, DIFF U UNCOUPLED, FIXED E

MIX WT = 0.05

YR RANGE =0.511E−01

A RANGE =0.110E+01

INT LOOP = 10

A MIN =0.205E+01

E MAX = 0.120

Figure 13.5

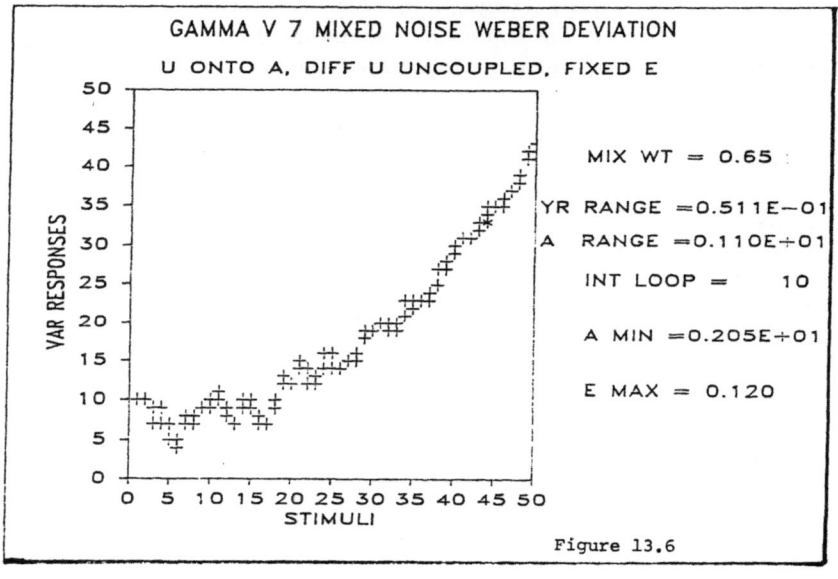

GAMMA V 7 MIXED NOISE WEBER DEVIATION

U ONTO A, DIFF U UNCOUPLED, FIXED E

MIX WT = 0.65

YR RANGE =0.511E−01

A RANGE =0.110E+01

INT LOOP = 10

A MIN =0.205E+01

E MAX = 0.120

Figure 13.6

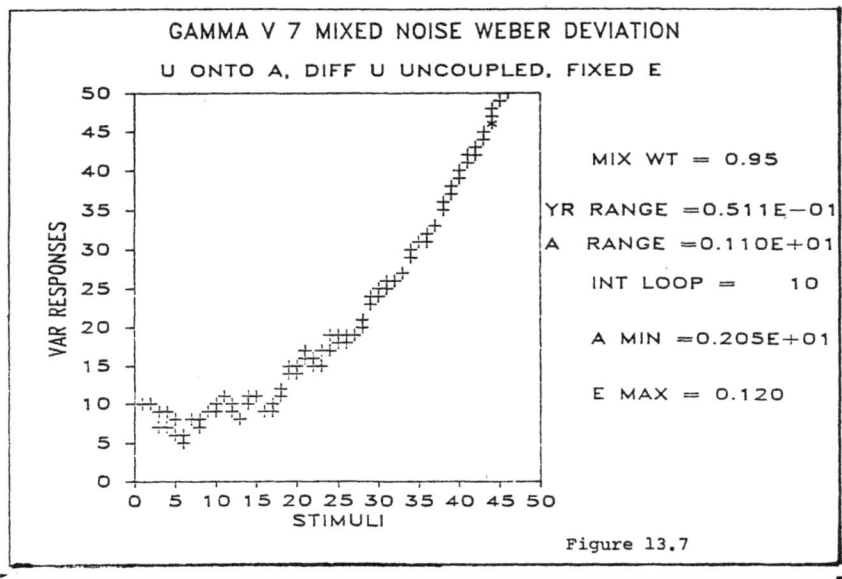

Figure 13.7

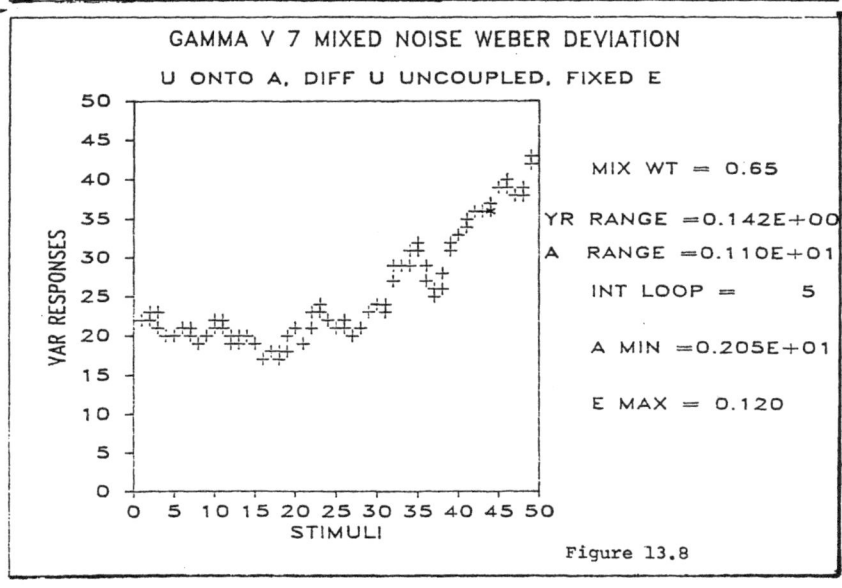

Figure 13.8

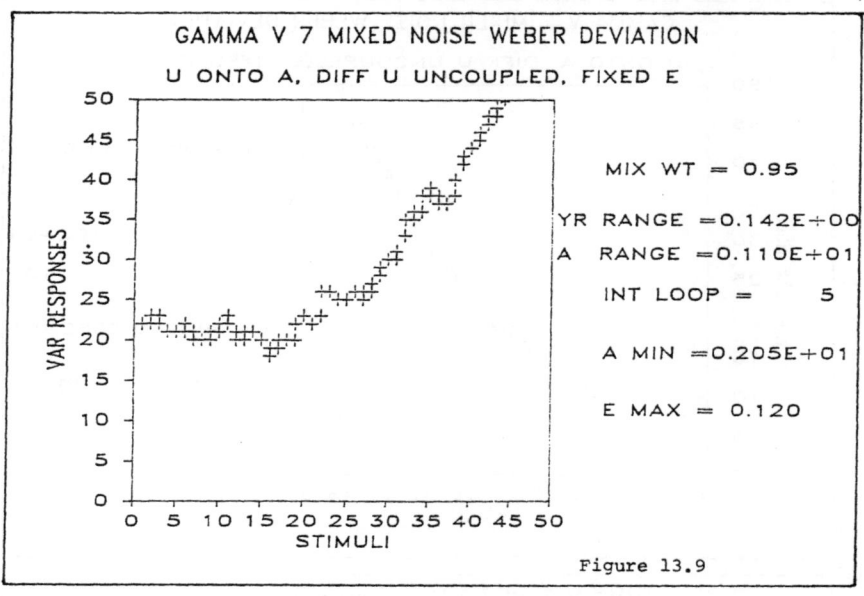

GAMMA V 7 MIXED NOISE WEBER DEVIATION

U ONTO A, DIFF U UNCOUPLED, FIXED E

MIX WT = 0.95

YR RANGE = 0.142E+00

A RANGE = 0.110E+01

INT LOOP = 5

A MIN = 0.205E+01

E MAX = 0.120

Figure 13.9

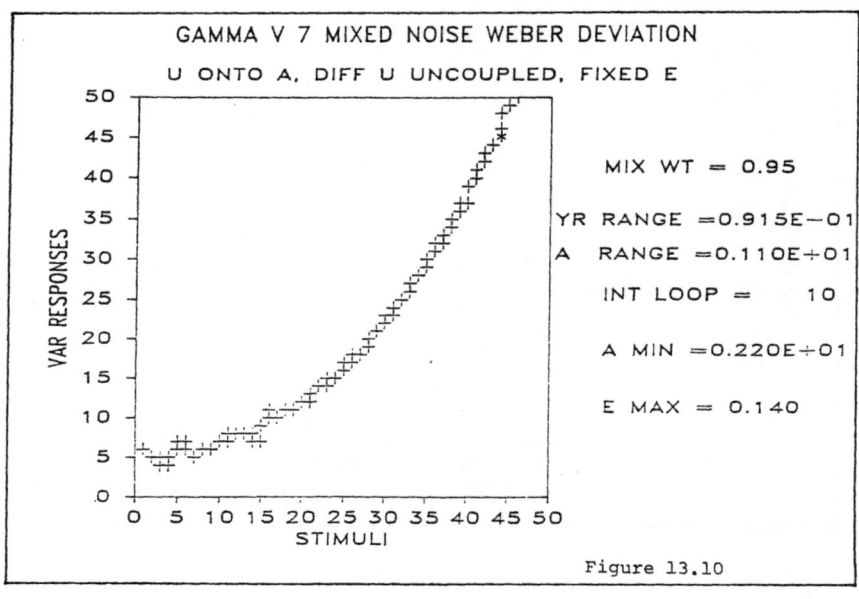

GAMMA V 7 MIXED NOISE WEBER DEVIATION

U ONTO A, DIFF U UNCOUPLED, FIXED E

MIX WT = 0.95

YR RANGE = 0.915E−01

A RANGE = 0.110E+01

INT LOOP = 10

A MIN = 0.220E+01

E MAX = 0.140

Figure 13.10

14 Résumé

In a sense the previous chapter has brought us full circle, by coming back to a problem that was already raising its head when Weber published **De Tactu** in 1834. A basic law, one of the few simple results in sensory processes, breaks down at extremes, and the extremes are readily encountered in a laboratory with versatile apparatus. There are four choices in science when this happens; make the theory more complicated, or scrap it (which needs courage), or start from another theory and show that odd things happen under limiting conditions, or write off the deviations as observational error. The fourth is quite popular, but here the third is to be preferred. The problem is, determining what constitutes limiting conditions, and how to represent them without proliferating parameters until the degrees of freedom almost match the number of anomalies originally found, as was noted at the end of chapter 11.

Until recently systems that showed discontinuities and a tendency to become turbulent were just intractable analytically. Experiments in psychology were designed to stabilize and simplify conditions until behaviour could, with some ruthless averaging of trends, and in some cases collecting thousands of responses, asymptotically converge on a static representation equation. In the words of Oscar Wilde, "Nature learns to imitate Art". The behaviour when subjects didn't conform to a model could, in the framework of such a traditon, always be interpreted as the consequences of transitory interruptions, or fluctuations in central variables that suddenly became perversely active, and so on. The fact that one can label such innumerable central variables with the vocabulary of either commonsense folkpsychology, or abstract decision theory, at choice, hardly satisfies William of Ockham's canon that explanatory entities should not be multiplied *without need.*

The position taken here is that a description of sensory behaviour, however accurate, that only describes behaviour in circumstances of constrained stability, achieved in an artificial stimulus-deprived context, is no description of any generality. All that can happen in warming-up, under very weak stimulation, in fatigue, during abrupt changes of stimulus levels, and with brain damage or drug-induced lesions, is all the same system under different conditions of input loading and rates of information processing. Ideally one system model should cope with all this, somehow, right from the first assumptions about its potential dynamics.

The existence of new developments in systems theory is something that should encourage psychophysicists to stop piling up results under induced stable conditions, which mask the very dynamics which create stability, and to look for, indeed to cherish, irregularities. Ohms law is very useful for designing circuits, but it doesn't describe what happens in the first few milliseconds when I switch on the light in my study. Nonlinear systems are no longer inaccessible, because we can simulate what cannot be treated analytically. But to know what to tell a computer to do requires some prior knowledge of equations like those in Chapter 2. It should be interesting to a psychologist that results in this area of mathematics were treated with some incredulity initially, and even if accepted for publication their full implications were not seen for years.

It may be said without too much irony that 98% of psychophysics assumes that there is only one sort of noise, and that it is normally distributed. With a lot of second-order effects, independently refusing to come under experimental control, that is plausible. But dynamic systems which do not function like random processes can mimic them quite well, and a different body of statistical analyses is needed to disentangle chaos from randomness. The "signature" of dynamic systems is typically one in which quite different sorts of behaviour arise, some quite compatible with a linear representation, and others anything but. It is not one sort of behaviour which is somehow more true or representative than another, but the total repertoire of such behaviour which the system can exhibit, and the conditions under which it jumps from one to another, that have to be teased out from the data. This holds both for the mathematical models in which we are interested, as in Chapter 2, and their analogies (if they are analogies) in observable behaviour.

The distinction has to be made between the behaviour of an equation, as in Chapter 2, and the total behaviour of a system, as in Chapter 3. A system model is a bridge between the curious dynamics of the algebra and the equally curious human behavioural dynamics we can observe. Modelling is a sort of constrained matching, a theme which recurs and surfaces, for example, in Chapter 9. Because both system models and human behaviour have many more properties, from the perspective of sequential analysis, than just a stimulus-response plot, many more tests of correspondence between theory and data are both possible and necessary. Dynamic processes do not give up their secrets readily to a pocket calculator with a stored regression program. For this reason, the preponderance of psychophysical studies that, painstakingly, merely report stimulus-response plots, with or without honest error variance bars on the graphs, are of little use except for designing spectacles, soft drinks, or acoustic tiles. These are worthy activities, but they are not the present concern. Additional data properties that can be computed, from Γ simulations, but not from static models, are noted in Chapters 4, 8, 9 and 10. Because we demand these extra properties of a model, the model itself is more vulnerable to refutation when it is wrong.

Repeatedly there have been references, in what may be surprising points in the argument, to neurophysiology and to clinical psychology. This does not imply that the author is a strong reductionist, though it does reveal a bias due to having acted at times as a consultant in hospital work on some defects of the chemical senses. There is a deeper reason, and that is, that using [2.2] leads to predictions that if a sensory system is overloaded, though not to the point of tissue damage, then it may lock onto different dynamics from those exhibited in the usual functional range of stimulation to which the organism adapts. What is the usual functional range is something labile, as considered in Chapter 12, and the presence of sensory illusions and hallucinations in epilepsy and like conditions should be seen as a clue to considering that the normal system is doing exactly what it would be expected to do if it were pushed too far.

There are two tentative interpretations made recurrently, from Chapter 3 onwards. First, that e represents in some way the sensitivity of the system to rates of change of inputs in time, sensation being viewed as dynamic tracking or, which is equivalent, as a return to stability after destabilization. Second, that the imaginary component of Y_{ob}, is a reflection of a sort of background noise intrinsic to the recursive loop process, but not necessarily the only sort of noise. It is possible to superimpose external random noise on internal deterministic noise, and the two can interplay. This topic has attracted, very properly, attention in theoretical physics and should certainly be rigorously analysed in biological systems. It is a topic which can be explored by analyses in the frequency domain, because the deterministic noise may have, characteristically, what is called a mixed $1/f$ and cycle frequency spectrum, nothing in appearance like white noise, with which it can mix. Let us be aphoristic; psychophysics is not about the soul responding to the music of the universe, as Fechner believed passionately, but about twitches of a mass of protoplasm trying to navigate its way through an unfeeling and only partly predictable world.

Instead of pushing these interpretations dogmatically, the case is allowed to build up in the context of a range of psychophysical problems that admit of a coherent analysis if such interpretations are considered. Take any one of the phenomena in Chapters 4 and 9 through 13 on its own, and the case is not convincing, but collectively one interpretation appears to make simple and consistent sense. There is still the problem of why this is so; there is no reason why parameters in a dynamic model of an internal process (internal to the system viewed as a black box) should have any commonsense intuitive interpretation. If we have acquired a lot of experience of real nonlinear systems, and also of peculiar differential equations, then we can spot analogies, but this is not commonsense behaviour on the part of the psychologist. The labels that theorists stick on psychological constructs or intervening variables are often words that have been around for centuries, which betrays their roots in conscious phenomenology. Science sometimes advances better by using, as building blocks, nouns and verbs which label things we could never experience through the senses.

In the preface the reader is warned that this present exercise is formative; there are no finalities in science, but sometimes there are better ways of looking at things. There are serious outstanding problems, of which five need mention. They could all be attacked with the present systems theory weapons, but they haven't been in these lectures.

(1) There is probably feedback between the final responses which the subject makes or utters, in actions, numerical symbols, or choice of labels, and the range of a and e values in the recursive loop. That is, the characteristics of the nonlinear recursion are constrained by a later mapping of its output, which has been left as Y_{obs}, and how this is used to generate observable behaviour. This comment goes for any psychophysical model, and is not an idea usually pursued except in studies of scaling *per se*. Nothing has been said at all about this aspect, but the question lurks within frame of reference ideas, as in Chapter 12; do a and e change as a function of U, or as a function of feedback comparing Y_{obs} and some reference level, say Y_{ref} ? Is there any way in which these two alternatives lead to different predictions about the sequential properties of long response records ?

(2) Nothing is in the model to represent sensory quality, only intensity, and the defence that there is nothing in a stimulus-response equation like the familiar $\Psi = a \cdot \Phi^b$ (the so-called power law relating physical activity levels Φ to sensory intensity Ψ) to encode quality, except to claim a unique value of b, is no defence as soon as we want some compatibility, as noted in Chapter 3, with a neural substrate. By seeking a common model for a unitary sensory pathway, only mediating intensity, the problem of sensory quality is pushed out to the periphery, or up to the cortex. Sensation of intensity without attendant identifiable quality (colourlessness, uniden-

tifiable odours, white noise with no pitch) does occur, and so a need for additional information to be collected or generated exists, quite apart from the transmission of intensity information as modelled here.

(3) There is nothing about reaction time, or response latency, in Γ, even though η represents the duration of internal processing. As η is only one segment in a pathway, and processes before and after the loop must have their own execution times, unless it is assumed that these other times are constant, or functionally linked in a stationary way to $\{a, e, \eta, W\}$, there is nothing to be said.

Attempts have been made to extend traditional and SDT models in psychophysics to give a basis for predicting observable reaction times. Perhaps the most successful rest on treating a response as a 'walk' towards an absorbing barrier; the 'walking time' having a distribution derivable from a theory of ergodic processes. Again, one could extend Γ here with ancillary assumptions, or one could, for example, explore the consequences of making η inversely proportional to $(a - a_{min})$. Actually, this has been tried out, with rather opaque results.

(4) Biological processes consume energy in order to function; there are limits on their rate of performance, and they do not go on for ever. This, again, is a feature that does not get incorporated in traditional models of sensory intensity; instead additional models of adaptation and recovery are needed. The special complication here arises in the prediction from Γ that very high inputs lead to an explosion in outputs, beyond chaos which is activity within bounds. This is not quite what happens in real life, people die from over-stimulation, and it is relatively easy to constrain the rate of the system, as we have done in Γ V 1, so that a very high input $U \mapsto a$ would almost immediately exhaust the system and push Y back to the baseline Y_0. It remains to be seen if such boundary conditions are enough to simulate the temporal course of outputs, for a series of a fixed input value U_{const}, which looks like adaptation.

(5) Even if qualitative encoding is falsely assumed to be present in a sensory pathway, the problem remains of how intensity information can be transmitted when there are two or more pathways excited at the same time. The special case in Chapter 13 is not enough, for it only draws on two components in the same channel[1]. Our own work on this problem,

---

[1] The problem with extending a model to cover the case where more than one quality is being transmitted at the same time, is that not only must more parameters be introduced to deal with cross-coupling between qualities, but that the structure of the recursion becomes one level of complexity worse. The mathematics in Chapter 2 are largely original, and more uncharted areas would need to be explored from an even more intricate starting point.

empirically, has been on odour mixtures. From a knowledge of the peculiar neurophysiology of the chemical senses, as well as from sensory experience, there is no good case for claiming that colours, smells, tastes and sounds all mix in the same way. For example, red and yellow can make orange, but musk and rose make a musky rose. There is a huge literature, and strong quantitative theory, on colour mixing, but relatively little on odour mixtures, reflecting both sociocultural priorities and relative difficulties of experimentation.

It is suggested that mixing sensations, in the framework of the Γ model, is something that happens by interaction upstream from the recursive loop, and that interaction is precisely that part of psychophysics which is unique and hence different for each sensory modality. Common laws of sensory intensity can hold for unidimensional stimulation, but not for mixing processes. But, as has been reiterated, nonlinear dynamics are not accessible to naïve intuition, cortical activity uses repeatedly the same bases of network operations and neurostructures to activate very diverse sensory phenomena, and so this conjecture could be overthrown by a more subtle sort of modelling than the one loop used here.

What it has been possible to show is that the issues called A1 to A19 in Chapter 2 are interrelated and not disparate, and that they are exactly what can occur in simple, self-stabilising, nonlinear systems if performance is demanded of an organism over a wide range of conditions. Self-stabilising combined with response sensitivity is a condition for biological survival, and the price of this combination is curious dynamics.

Subject Index